EXPLORING THE GEOMETRY OF NATURE

Computer Modeling of Chaos, Fractals Cellular Automata and Neural Networks

Edward Rietman

WINDCREST

To the people of Vietnam and Cambodia.

Published by **Windcrest Books**
FIRST EDITION/THIRD PRINTING

Library of Congress Cataloging-in-Publication Data

Rietman, Ed.
    Exploring the geometry of nature : computer modeling of chaos.
fractals, cellular automata, and neural networks / by Edward
Rietman.
      p.   cm.
    Bibliography: p.
    Includes index.
    ISBN 0-8306-9137-5   ISBN 0-8306-3137-2 (pbk.)
    1. Chaotic behavior in systems--Mathematical models. 2. Fractals-
-Mathematical models. 3. Cellular automata--Mathematical models.
    4. Neural circuitry--Mathematical models. I. Title.
    QA845.R48 1988
003-dc19   88-24878
                                        CIP

TAB BOOKS Inc. offers software for sale. For information and a catalog, please contact TAB Software Department, Blue Ridge Summit, PA 17294-0850.

Questions regarding the content of this book should be addressed to:
Windcrest Books
Division of TAB BOOKS Inc.
Blue Ridge Summit, PA 17294-0850

Ron Powers: Director of Acquisitions
Marianne Krcma: Technical Editor
Jaclyn B. Saunders: Series Design

Contents

List of Programs

All programs were written on an AT&T 6300 (on IBM clone), with MS-DOS 3.1, GWBASIC, and an 8087 coprocessor.

Preface

This is a first-level introduction to computer modeling of chaos and self-organizing systems. I started modeling these phenomena in 1980 on a TRS-80 model III computer. Then, in 1985, I upgraded to a PC clone with a high-resolution monitor. I spent so much time modeling chaos I knew an introductory book on the subject was needed.

I have attempted to cover in a simple way what I feel is important and interesting in the field of chaos. I have, by no means, covered the entire field. That couldn't be done in an introductory book. It would require several volumes. The level of this book requires a good physics and math background, with some chemistry, biology, and computer programming skills.

Acknowledgment

There are many people I would like to thank. First I would like to thank Cody Stumpo for putting up with my low blood sugar. I thank Peter Littlewood for reviewing several chapters. I thank Matthew Marcus, Robert Barns, and Sam Martin for helpful conversations. I would also like to thank Ron Powers and the editorial staff of TAB BOOKS Inc.

I thank my best friend, my lover, my personal editor, my wife—Suzanne Harvey.

Introduction

The science of chaos has become a significant study that has attracted the attention of scientists in many disciplines ranging from social science and biological science to chemical and physical sciences. This book describes examples from many of these fields. You are expected to have a good physics and/or mathematics background for understanding the contents. If you do not have this background, you can still gain much from using the programs and observing the dynamical maps plotted on a computer.

The bulk of the programs should work on almost any computer running BASIC. I use an IBM XT clone with an 8087 coprocessor chip. I wrote the programs to generate a data file that another program must read and plot. By using this approach, the bulk of the programs can be used on many types of computers by changing the appropriate section of each program that writes to a data file. Then the data file is read and plotted by a separate program. I include a data-file reading and plotting program, but, unlike the other programs, it might require extensive modification to run on a non-clone computer.

In CHAPTER 1, I discuss the mathematical methods used to study chaos. These include graph theory, set theory, and differential equations. The chapter includes a computer program to solve systems of differential equations. CHAPTER 2 covers chaotic maps in one dimension. These are known as *iterated maps*. Included are several computer programs to generate data files of iterated maps, and a plotting program for the data files.

In Chapter 3, I discuss strange attractors in understandable terms and give several examples of them. The chapter also includes a few applications and a short catalog of strange attractors for the experimenter to spend time studying.

Chapter 4 covers cellular automata. I discuss what cellular automata are, applications, and rules for generating them. The chapter includes several computer programs for demonstrating cellular automata, and a brief description of some applications in the study of chaos and self-organization. Chapter 5 deals with neural dynamics, in which I describe an artificial neural network that is a type of threshold cellular automata. This chapter includes a computer program to model neural networks, and a discussion of neural dynamics and chaos in neural systems.

Chapter 6 is a short introduction to fractals and Julia sets. The bulk of the chapter is a discussion of the algorithm for the computer generation of Julia sets and the Mandelbrot set. Also included in this chapter is a computer program for Julia sets and a short introduction to some of the applications of fractals.

Chapter 7 is about self-organizing systems, including a short review of the diverse areas in which self-organization is observed, and an introduction to the mathematics of the phenomenon.

1
Mathematical Techniques

This first chapter is a discussion of some of the mathematical techniques used to study chaos and self-organizing systems. It begins with a review of the elements of set theory and graph theory. Then, differential equations and difference equations are reviewed. The chapter concludes with a computer program to solve systems of differential equations using a Taylor series expansion.

SET THEORY

This section on set theory is little more than a review of the introductory aspects, including examples from dynamical systems that are similar to mappings of chaotic dynamical systems. This is included primarily so you can understand some of the terminology used in research journals and books on chaotic dynamical systems. Much of the following is similar to that found in *Set Theory: An Intuitive Approach,* by Lin and Lin (1974). Logical statements are symbolically represented througout this book by lowercase letters such as *p*, *q*, and *r*. These can be combined to form compound statements. There are only five common connectives, all of which are shown in Table 1-1.

Table 1-1

Connective Word	Connective Symbol
NOT	$\sim$
AND	$\wedge$
OR	$\vee$
IF . . . . THEN . . .	$\rightarrow$
. . . IF AND ONLY IF . . .	$\leftrightarrow$

Table 1-1 shows the common Boolean logic connectives familiar to computer programers. The phrase *if and only if* is sometimes written as *iff*. As stated above you can make compound statements with these connectives. For example if *p* is a statement, then ~*p* reads "not *p*," or the negative of *p*. Table 1-2 is a simple truth table for this example.

Table 1-2

p	~p
T	F
F	T

Another example is $p \wedge q$. This is read "*p* and *q*," or the conjunctive of *p* and *q*. This is an example of a compound statement. The truth table for this example is given in Table 1-3. A more complex statement truth table can also be constructed, such as that in Table 1-4.

Table 1-3

p	q	p∧q
T	T	T
T	F	F
F	T	F
F	F	F

Table 1-4

p	~p	p∨~p
T	F	T
F	T	T

For any given discussion concerning a set or group of objects, it is common to see a statement such as "for all *x* in the set . . . ." This is a *universal quantifier* and is symbolized as (∀*x*). Another common phrase is, "there exists at least one *x* such that . . . ." This is called an *existential quantifier* and is symbolized as (∃*x*). Now let's use these two definitions to make more complex statements.

Given a domain, *U*, which is a collection of objects under consideration, and a general statement, *p(x)* (called a *propositional predicate*), whose variable *x* ranges over *U*, we can make the following statement:

$$(\forall x)(P(x))$$

This says that for all x in U, the statement p(x) about x is true. Another example is shown below:

$$(\exists x)(P(x))$$

This means that there exists at least one x in U such that p(x) is true. In summary, the statement f(x) = 0 *for all* x is just the same as this:

$$(\forall x)(f(x)) = 0$$

Set Notation

A *set* is any collection of distinguishable objects, called *elements*. A set that contains only finitely many elements is called a *finite set;* an infinite set is one that is not a finite set. Sets are frequently designated by enclosing symbols representing their elements in braces. The empty set is called a *null set*, and is denoted by the symbol {∅}. If a is an element of set A, we write a ∈ A, which is read "a is an element of A," or a belongs to A. Similarly, b ∉ A means that b is not an element of A.

Two identical sets are represented as follows:

$$(\forall x)[(x \in A) \leftrightarrow (x \in B)]$$

The order of elements of a set is irrelevant. Set {a,b,c} is the same as {b,c,a} or {c,b,a}.

Another important concept is *subsets*. If every element of set A is also contained in set B, then A is a subset of B. In symbols, this is written A ⊆ B or B ⊇ A. Of course, if A is a subset of B, then B is a superset of A.

$$(A \subseteq B) \equiv (\forall x)[(x \in A) \rightarrow (x \in B)]$$

Naturally, every set is a subset and superset of itself. When A ⊆ B and A ≠ B, we write A ⊂ B or B ⊃ A, which reads "A is a proper subset of B," or "B is a proper superset of A." The empty set is a subset of every set.

$$(x \in \emptyset) \rightarrow (x \in A)$$

Set Builder Notation

To every set A and to every statement p(x) about x ∈ A, there exists a set

$$\{x \in A \,|\, P(x)\}$$

whose elements are those elements x of A for which the statement $p(x)$ is true. The statement

$$\{x \in A \,|\, P(x)\}$$

is read "the set of all x in A such that $p(x)$ is true." This notation is the *set builder notation*.

In arithmetic addition, multiplication and subtraction are operations on numbers. Analogous operations can be performed on sets. The *union* of two sets is represented by $A \cup B$. This results in a set of all elements x, such that x belongs to at least one of the two sets A and B. That is

$$x \in A \cup B$$

if and only if

$$(x \in A) \vee (x \in B)$$

The *intersection* of two sets A and B is represented by $A \cap B$. It results in a set of all elements x which belong to both A and B. In symbols,

$$A \cap B = \{x \,|\, (x \in A) \wedge (x \in B)\}$$

or

$$\{x \in A \,|\, x \in B\}$$

As an example, let $A = \{1,2,3,4\}$ and $B = \{3,4,5\}$. Then

$$A \cup B = \{1,2,3,4,5\}$$
$$A \cap B = \{3,4\}$$

The *complement* of B in A is the set $A - B$, symbolized by

$$A - B = \{x \in A \,|\, x \notin B\}$$

As an example, let $A = \{a,b,c,d\}$ and $B = \{c,d,e,f\}$. Then,

$$A - B = \{a,b,c,d\} - \{c,d,e,f\} = \{a,b\}$$

$$A - (A \cup B) = \{a,b,c,d\} - \{c,d\} = \{a,b\}$$

You should note that

$$A - B \not\equiv B - A$$

Sets in Dynamical Systems Theory

Let us now look at some examples of set theory notation used in nonlinear dynamics system theory. In one-dimensional iterated maps of chaotic dynamical systems, you will often see the following relation:

$$x_{n+1} = f(x_n), \quad x_n \in [0,1], \quad n = 0, 1, 2, \cdots$$

This is a difference equation of a unimodal map, i.e., the mapping is contained in the unit interval.

Another example from dynamical systems theory is the Cantor Middle-Thirds set. Start with the unit interval and remove the middle third. Next, remove from the two middle thirds from what remains. Continue removing middle thirds in this fashion. At the nth stage, 2^n open intervals are removed. Figure 1-1 shows a schematic of this procedure. The Cantor Middle-Thirds set is an example of a *fractal*. A fractal is a set that is self-similar under magnification. Fractals are found in Chapters 2 and 3, and are covered more fully in Chapter 6.

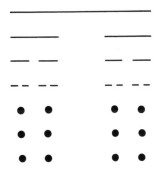

Fig. 1-1.   Construction of a Cantor set.

GRAPH THEORY

This section covers a few points of graph theory. This section is an introduction and/or review so you will understand the terminology used in research journals and advanced books. An example of graph theory as it applies to neural networks and parallel processing is given later in this section. Much of this section is similar to *The Theory of Graphs: A Basis for Network Theory*, by Maxwell and Reed (1971).

A *graph* is a set of points. The points are called *vertices*, and they are connected by lines called *edges*. These graphs have no properties other than visual. Graph theory is a study of the interrelationships between vertices and edges. Graphs have many applications, including game theory, networks, flow diagrams, molecular structure, and family trees. Network applications are used in the study of iterated maps.

For a network containing e elements, it is necessary to solve a system of 2e equations. Later in this section you will learn how to construct a graph from a matrix, but first, a few basic definitions are necessary. A subgraph, G_s of a graph G is a subset of the set G.

$$G_s \subset G$$

Connected graphs are called *circuits* if each vertex is of degree two. In other words, each element is a two-terminal device. Several circuit examples are given in Fig. 1-2.

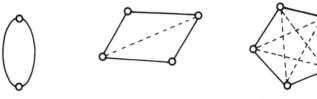

Fig. 1-2. Examples of circuit graphs.

In the three examples pictured in Fig. 1-2, the subgraphs are also circuits shown by dotted lines. Circuits are distinguished by several properties: A circuit contains no end elements. It contains only interior vertices. A circuit contains at least two elements, and is always a connected planar graph.

Besides the circuit, another important graph is the *tree*. A tree is a subgraph of a point, *P*, such that it contains no circuits, is connected, and contains all the vertices of point, *P*. Figure 1-3 is an example of a tree

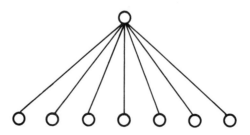

Fig. 1-3. Example of a tree graph.

graph. Tree graphs are models for interconnection of parallel processing computers. They are also used in algorithms for artificial intelligence solutions to games.

Constructing a Graph From a Matrix

A *directed graph* is a graph in which an arrowhead is assigned to each element of the graph. Given the matrix M, you can construct a directed graph. This matrix is given as follows:

$$M = \begin{bmatrix} 0 & 1 & 0 & 0 & 0 \\ 0 & 0 & 1 & 0 & 0 \\ 0 & 0 & 0 & 1 & 0 \\ 0 & 0 & 0 & 0 & 1 \\ 1 & 0 & 0 & 0 & 0 \end{bmatrix}$$

This matrix is described by the following mapping:

$M(1,1) = 0 \quad M(2,1) = 0 \quad M(3,1) = 0 \quad M(4,1) = 0 \quad M(5,1) = 1$

$M(1,2) = 1 \quad M(2,2) = 0 \quad M(3,2) = 0 \quad M(4,2) = 0 \quad M(5,2) = 0$

$M(1,3) = 0 \quad M(2,3) = 1 \quad M(3,3) = 0 \quad M(4,3) = 0 \quad M(5,3) = 0$

$M(1,4) = 0 \quad M(2,4) = 0 \quad M(3,4) = 1 \quad M(4,4) = 0 \quad M(5,4) = 0$

$M(1,5) = 0 \quad M(2,5) = 0 \quad M(3,5) = 0 \quad M(4,5) = 1 \quad M(5,5) = 1$

From this mapping you can construct the two equivalent graphs shown in Fig. 1-4. You should note that the spatial arrangement of the points is irrelevent.

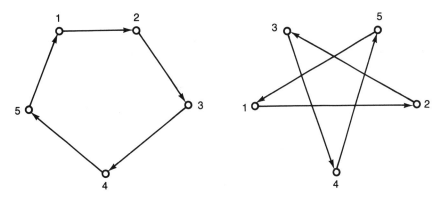

Fig. 1-4. Example of two equivalent graphs.

Discrete Iterations

Another interesting example of graph theory is in discrete iterations. Dewdney (1986) wrote an introduction to this model and Robert (1986) has gone into far more detail on discrete iterations. To illustrate discrete operations, pick any number at random between 0 and 99. Find its square, take the last two digits of this result, and square this number. Repeat this process and eventually you will encounter a number you have already encountered. As an example, take 81 and square it:

$$81^2 = 6561$$

$$61^2 = 3721$$

$$21^2 = 441$$

$$41^2 = 1681$$

$$81^2 = 6561$$

This leads to a cycle of period four. From this result, you could produce the graph shown in Fig. 1-5.

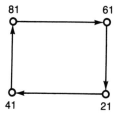

Fig. 1-5.   Graphical example of a period-four cycle.

DIFFERENTIAL EQUATIONS

The opening few paragraphs of this section provide a review of calculus, followed by a discussion of differential equations and difference equations. The final part of this chapter describes an algorithm and computer program to solve systems of differential equations using a Taylor series expansion.

Let's begin with the integral, because it is easy to grasp graphically as an area under a curve. Given a curve such as the one shown in

Fig. 1-6, describing the function $v(t)$, the area under the curve is given by

$$I = \int_0^t V\,dt$$

where dt is an infinitesimally small interval of time. The integral symbol $\int$ is known as a "lazy s," and represents the summation of the product of $v\,dt$ from zero to t_1.

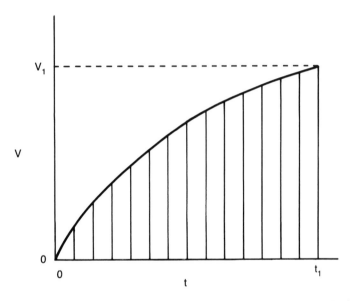

Fig. 1-6. Plot of a simple function. The area under this curve is the integral of the function from zero to t_1.

Differentiation is the opposite of integration. In differentiation, a small change in v, represented by dv is divided by an infinitesimally small time interval, dt. The result $\dfrac{dv}{dt}$, is known as the derivative of v with respect to t. These infinitesimally small changes can be represented as small changes in v with respect to a small change in t:

$$\frac{dv}{dt} \approx \frac{\Delta v}{\Delta t}$$

The derivative is given as the slope of the curve v(t) evaluated at the point of interest. This is an important concept that is used in the solution of differential equations and in evaluating the critical properties of chaotic systems.

An example of a simple differential equation for a series circuit shown in Fig. 1-7.

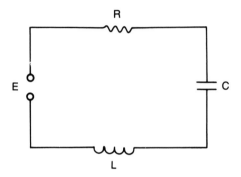

R

E

C

L

Fig. 1-7.   Simple electrical circuit for the differential equation example.

For resistance R, capacitance C, inductance L, and voltage source E, the current flowing around the circuit I(t) at time t is given by

$$L\frac{dI}{dt} + RI + \frac{q}{C} = E$$

where RI is the voltage across R, and the voltage across C is given by q/C. The voltage across the inductance L is given by $L(dI/dt)$. If you differentiate this equation with respect to time and substitute $dq/dt = I$, you get the following:

$$L\frac{d^2I}{dt^2} + R\frac{dI}{dt} + \frac{I}{c} = \frac{dE}{dt}$$

Now that I have reviewed the introductory concepts of calculus and shown how differential equations are built up from derivatives, I will show how to solve a differential equation.

For a simple example of solving a differential equation, let's start with what is known as a *first-order differential equation*. Consider the following simple equation:

$$\frac{dy}{dx} = \sin(x)$$

This can be solved by separation of variables, as follows:

$$dy = \sin(x)dx$$

$$\int dy = \int \sin(x)dx$$

$$y = -\cos(x) + c$$

where c is the constant of integration.

An example of a nonlinear differential equation in dynamics is the motion of a damped pendulum. The equation for this system is the following:

$$ml^2\frac{d^2\theta}{dt^2} + cl\frac{d\theta}{dt} + mgl\sin(\theta) = 0$$

From Fig. 1-8, the angle θ is the angular displacement of the pendulum from the vertical.

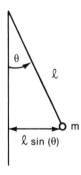

Fig. 1-8.   Example of a simple pendulum.

The damping constant is given by c and is always greater than zero. The mass is given by m and the length and gravitational constant are given by l and g, respectively. This equation can be solved by substitution:

$$\text{Let} \quad x = 0$$

$$y = \frac{d\theta}{dt}$$

$$\text{Then} \quad \frac{dx}{dt} = y$$

$$\frac{dy}{dt} = -\frac{g}{l} \sin (x) - \frac{c}{ml} y$$

This gives a system of differential equations. This is an example of a two-dimensional system. Later, in Chapter 3, you will see some three-dimensional systems that result in strange attractors. Rather than actually solve this system analytically, I will now introduce a computer algorithm for the solution.

ALGORITHMS FOR SOLVING DIFFERENTIAL EQUATIONS

There are many books that discuss computer solutions to differential equations. There is an excellent chapter in *Elementary Differential Equations* (Boyce and DiPrima, 1977). Shoup (1983), has written an excellent book, *Numerical Methods for the Personal Computer*, and Danby (1985) has written a small book with hundreds of examples of differential equations for solving with a personal computer called *Computing Applications to Differential Equations*. I should also mention the advanced book by Potter (1973) *Computational Physics*, which is devoted to computer modelling in physics.

The simplest method for the solution of differential equations is a one-step method known as Euler's method. This method is outlined in Fig. 1-9.

The principle of the method involves a Taylor series expansion of the following form

$$y(x_o + h) = y(x_o) + hy'(x_o) + \tfrac{1}{2}h^2y''(x_o) + \cdots$$

Taking a small step h from the initial value, we can see that if h is indeed small, then h^2 is even smaller, and so the equation can be approximated by

$$y(x_o + h) = y(x_o) + hy'(x_o)$$

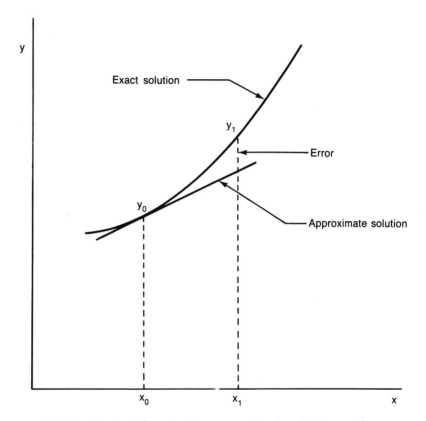

Fig. 1-9. Graphical example of the error produced by the Euler method.

This can be written as a difference equation:

$$y_{n+1} = y_n + hf(x_n, y_n), \quad n = 1, 2, \cdots$$

In the Euler method, the slope of the curve for the initial value is exact. The slope changes at the step value $x_0 + h$, giving an error. In the modified Euler method, a better solution is found by taking an average value of the derivatives at the beginning and end of the interval. This average value is then used to calculate the derivative at the end of the interval.

There are other methods to solve differential equations; these include the Ruge-Kutta method, Gill's method, Milne's method and the Adams-Bashforth method. These methods are not discussed here. If you are interested you should check out Shoup, or Boyce and DiPrima. These other methods can give very high accuracy with good speed for the computer solution. I selected the simple Euler method for the computer program because it is very easy to modify the program for a system of n equations, but the method is slow. Chapter 3 discusses the Lorenz system of three equations. This solution required ten hours computation time. The accuracy goes up as the step size goes down, but of course the CPU time increases.

COMPUTER SOLUTION OF DIFFERENTIAL EQUATIONS

The computer program described here to solve systems of differential equations is based on the damped pendulum example. The system, as derived earlier in this chapter, is as follows:

$$\frac{dx}{dt} = y$$

$$\frac{dy}{dt} = -\frac{g}{l} \sin (x) - \frac{c}{ml} y$$

The program was written in BASIC and should run on any system. This program, SDEQ1, solves a system of differential equations using the Euler method.

SDEQ1

```
10 REM DEFINE DX/DT=D2=F(T,X,Y,Z) IN LINE 130
20 REM DEFINE DY/DT=D1=F(T,X,Y,Z) IN LINE 140
30 INPUT "INPUT INITIAL AND FINAL VALUES OF T ";T1,T2
40 INPUT "INPUT DELTA T ";D
50 INPUT "INPUT INITIAL CONDITIONS X,Y ";X,Y
60 INPUT "INPUT NUMBER OF CALCULATIONS FOR EACH DELTA T ";N
70 INPUT "INPUT FILE NAME ";FILE$
80 OPEN "O",#1,FILE$
90 FOR T9=T1 TO T2 STEP D
100 PRINT T9,X,Y
110 PRINT #1,X,Y
120 FOR T=T9 TO T9+D STEP D/N
130 D1=Y
140 D2=-6*SIN(X)-5*Y
150 X=X+D1*D/N
160 Y=Y+D2*D/N
170 NEXT T
180 NEXT T9
190 CLOSE #1
200 END
```

Let's look at the program, line by line. Line 30 allows the user to input the initial and final time, T_1 and T_2. In line 40 the user enters the time increment. This has the variable name, D. I usually select a value of about 0.1 for the time increment. Smaller values can be used, but computation time increases and the number of data points generated increases rapidly. In line 50 the initial conditions are entered. These are the initial conditions for the x and y values.

In line 60 the number of calculations for each time increment is entered. I usually enter 50. Some explanation might be needed as to what this number represents. Earlier in this chapter, I showed that the Euler method can be written as a truncated difference equation derived from a Taylor series expansion. The difference equations for the two equation system in this example can be written as follows:

$$x_{n+1} = x_n + hf(x_n, y_n, t_n)$$

$$Y_{n+1} = y_n + hg(x_n, y_n, t_n)$$

In this system, the function $f(x_n, y_n, t_n)$ is the derivative dy/dt, which in our example is

$$f(x_n, y_n, t_n) = \frac{dx}{dt} = y$$

Similarly, the function $g(x_n, y_n, t_n)$ is

$$g(x_n, y_n, t_n) = \frac{dy}{dt} = -\frac{g}{l}\sin(x) - \frac{c}{ml}y$$

The parameter h is directly related to the error. If h is very small, the error is also very small (but the computation time increases quickly). This h value is given by the relation

$$h = \frac{\Delta t}{N}$$

where Δt is the time increment (in the program this is named D) and N is the number of calculations per time increment.

To return to the line-by-line discussion, line 70 asks the user to enter a file name. The computed data points are stored in this file and then plotted or manipulated with a separate program. This will be discussed later in this chapter. Line 80 opens the file.

In line 90, the calculation begins. The loop is set up to increment from T_1 (the initial time) to the final time, T_2, in a step size D, or delta time. Line 100 prints the x,y values to the computer display, and line 110 prints these values to the file. The first time through the loop, the initial values are printed. In line 120, the calculation begins. A loop is started to calculate the derivative using the difference equation and incrementing the step size D/N, as defined previously. This is the time increment divided by the number of calculations per increment. Line 130 and 140 define the differential equation and the difference equation. The loops are repeated until the end. Then the file is closed in line 190 and the program ends.

Now let's run the program. First notice that the program is set up for the pendulum example. By changing the differential equations in lines 130 and 140, you could investigate a different system. For this system I selected the values of the constants as follows:

$$\frac{g}{l} = 6$$

$$\frac{c}{ml} = 5$$

After entering RUN, I selected the initial time as 0 and the final time as 100, with a time increment of 0.1. The initial condition I chose was the point (5,5) on the x,y plane. I selected the number of calculations between each time step to be 25. You can run the program yourself to generate the data file. Once you have the data file, you can then graph the data. My plot of this file for the pendulum example is given in Fig. 1-10.

Notice that the curve starts at the point $y = 3.0$ and quickly falls to $y = 0$ at $t = 4.2$. The quick relaxation of this pendulum is due to the damping coefficient, c. In the earlier example, c was divided by mass and length to give a new constant with a value of 5.

I would like to reiterate that this is a general-purpose program and can be easily modified to solve a system of n differential equations. This small program is used over and over in the other examples throughout this book without any further discussion of the solutions of differential equations or the algorithms.

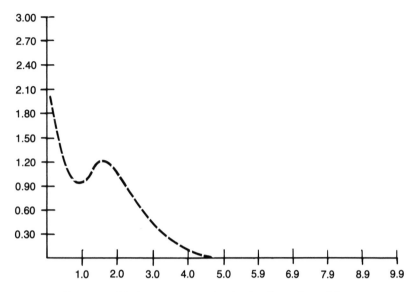

Fig. 1-10.   Displacement-time plot for the damped pendulum.

DATA-FILE PLOTTING PROGRAM

The following program is for an IBM PC clone. Most of the programs in this book, those that do calculations and/or generate data files, should work on any computer's version of BASIC with a little modification for file storage. This graphic program would require extensive modification for non-IBM clones. By making the majority of programs generate data files rather than plotting simultaneously, a larger number of readers can experiment with these dynamical systems. If you do not have an IBM PC clone, you will have to write your own graphics module.

The program PLOT1 can be used to plot two-dimensional data files, or the program can be modified to plot three-dimensional data files by selecting a two-dimensional slice through the three-dimensional space. I only discuss some of the program highlights here, rather than giving a line-by-line description.

PLOT1

```
10 CLS
20 INPUT "INPUT NUMBER OF POINTS ";NPTS
30 DIM X(5001),Y(5001)
40 CLS
```

```
50 INPUT "what is the name of the disk file you want to
   plot";FILENAME$
60 OPEN "I",2,FILENAME$
70 FOR I=1 TO NPTS
80 IF EOF(2) THEN 110
90 INPUT#2,X(I),Y(I)
100 NEXT I
110 NPTS=I-1
120 XMAX=-1E+20 :XMIN=-XMAX
130 YMAX=-1E+20
140 YMIN=-YMAX
150 FOR I=1 TO NPTS
160 IF YMIN>Y(I) THEN YMIN=Y(I)
170 IF YMAX<Y(I) THEN YMAX=Y(I)
180 IF XMAX<X(I) THEN XMAX=X(I)
190 IF XMIN>X(I) THEN XMIN=X(I)
200 NEXT I
210 CLS
220 NXTIC=10:NYTIC=10
230 XMN=XMIN:XMX=XMAX:YMN=YMIN:YMX=YMAX
240 CLS
250 SCREEN 2
260 DSX=ABS(XMX-XMN):DSY=ABS(YMX-YMN)
270 SX=.1:SY=.1
280 AXMN=XMN-DSX*SX:AXMX=XMX+DSX*SX
290 AYMX=YMX+DSY*SY:AYMN=YMN-DSY*SY
300 WINDOW (AXMN,AYMN)-(AXMX,AYMX)
310 LINE (XMN,YMN)-(XMX,YMN)
320 LINE (XMN,YMN)-(XMN,YMX)
330 DXTIC=DSX*.02:DYTIC=DSY*.025
340 XTIC=DSX/NXTIC:YTIC=DSY/NYTIC
350 FOR I=1 TO NXTIC
360 XP=XMN+XTIC*I
370 LINE (XP,YMN)-(XP,YMN+DYTIC)
380 ROW=24
390 NEXT I
400 FOR I= 1 TO NYTIC
410 YP=YMN+I*YTIC
420 LINE (XMN,YP)-(XMN+DXTIC,YP)
430 NEXT I
440 FOR I=1 TO NPTS-1
450 J=I+1
460 IF Y(I)>YMX OR Y(J)>YMX OR Y(J)<YMN OR X(I)<XMN
    THEN 480
470 CIRCLE (X(I),Y(I)),0
480 NEXT I
490 FOR I=1 TO NXTIC
500 XP=XMN+XTIC*I
```

```
510 XC=PMAP(XP,0)
520 COL=INT(80*XC/640)-1
530 LOCATE 24,COL
540 PRINT USING "###.#"; XP;
550 NEXT I
560 FOR I=1 TO NYTIC
570 YP=YMN+I*YTIC
580 YR=PMAP(YP,1)
590 ROW=CINT(24*YR/199)+1
600 LOCATE ROW,1
610 PRINT USING "###.##"; YP
620 NEXT I
630 GOTO 630
```

After clearing the screen, the program prompts the user for the number of data points in the file and the file name. The file is then read into a matrix that has been dimensioned in line 30. After reading the entire file, the minimum and maximum value for both the abscissa and ordinate is found. The screen is then cleared in line 210, and axes and tick marks are drawn on the computer display. The data points from the matrix are then plotted on the display. After displaying the numerical values for the tick marks on the axes, the program enters an infinite loop to prevent the cursor from appearing on the display. The user can then press the print graphics keys for a hard copy of the graph.

2

One-Dimensional
Iterated Maps

Chapter 1 includes an equation for a one-dimensional iterated map on the interval. I used it then as an example of set theory notation. In this chapter, I describe a period-doubling route to chaos and define bifurcation. I also give extensive description of a map of the form

$$x_{n+1} = f(x_n), \quad x_n \in [0,1], \quad n = 0, 1, 2, \cdots$$

that is, a unimodal map. These maps are contained in the unit interval. At the end of this chapter there is a short list of systems that you can investigate on your personal computer. But why should you study these maps and what do they have to do with modelling chaos?

Chaos has been observed in many fields, including physical, biological, social, and chemical systems. For centuries, since Newton, physics has taken a reductionist approach. That is, a system has been broken down into subsystems or components and laws governing their behavior have been discovered and investigated. In the new science of deterministic chaos, a more holistic or synergistic approach is taken. Many questions require a synthesis rather than a reductionist approach. Some of these questions concern the evolution of biological systems and the principles underlying the operation of the brain. A question we will concern ourselves with throughout this book is: How can a system become chaotic? In other words, we will concern ourselves with the transition to turbulence.

At the outset, I'd like to mention the difference between *dissipative* and *nondissipative* systems. A system in which energy is dissipated by friction, or its analog, is called a dissipative system. The presence of this friction implies the existence of an *attractor*. An attractor is an asymptotic limit of the solution as time approaches infinity. Attractors can be limit cycles, single points, or simple oscillators. Sometimes, depending on the controllable quantity known as the *parameter*, the attractor becomes a *strange attractor*. These attractors show great sensitivity to the initial conditions, and have a fractal dimension. These are studied in more detail in Chapter 3.

THE LOGISTIC MAP

This section provides a detailed discussion of the iterated map known as the logistic map. It also introduces the route to chaos and bifurcation theory. Among other topics relevant to the map discussion, I will introduce a computer program to model one-dimensional maps.

There are a number of excellent discussions on one-dimensional maps, and in particular the logistic map. Devaney (1986) has an entire chapter on one-dimensional maps in *An Introduction to Chaotic Dynamical Systems*. Lauwerier (1986) has a chapter on one-dimensional maps published in *Chaos* edited by Holden (1986). There are a number of papers in Barenblatt, et al. (1983) and an important chapter in Berge, et al. (1984). Finally, I must mention *Iterated Maps on the Interval as Dynamical Systems*, an excellent book by Collet and Eckmann (1980) on iterated maps on the interval. Much of the original research on the logistic map was reported by May (1976) and Feigenbaum (1980), who discovered two universal constants to describe the transition to chaos.

The logistic map is a model of population dynamics. A very simple model is given by the following differential equation:

$$\frac{dx}{dt} = kx$$

For $k < 0$, this relation is undefined for population dynamics. This is the exponential growth or decay equation introduced to describe the population, x, of a single species. In this relation, t is the time and k is a constant. This differential equation models the system in continuous time. We will model the system in discrete time stages. Each time step could be the generation time, or perhaps a year in the life of the population.

The discrete model is given by

$$x_{n+1} = kx_n$$

This is still a naive model. A more accurate model is given by the differential equation

$$\frac{dx}{dt} = kx(L - x)$$

where L is a limiting value for the population. This keeps the growth/death in check, causing the population to stabilize. The analogous difference equation leads to very complicated dynamics and eventually to chaos, depending on the parameter, r. The analogous difference equation, after rescaling, is as follows:

$$x_{n+1} = rx_n(1 - x_n)$$

This is the logistic map that we will study on the unit interval:

$$x_{n+1} = rx_n(1 - x_n), \quad x_n \in [0,1], \quad n = 0, 1, 2, \cdots$$

It can be clearly seen from this equation that if x_n is greater than one, then x_{n+1} is negative, which is meaningless from a population dynamics view. That is why the analysis is confined to the interval $0 \leq x \leq 1$. The function then transforms any point on the unit interval into another point on the same interval.

ATTRACTOR POINTS AND BIFURCATIONS

I will now discuss the concept of attractor points, also called fixed points, or *bifurcations*. Figure 2-1 is a plot of the logistic map.

$$x_{n+1} = Ax_n(1 - x_n)$$

The parabolic-shaped curve is the function. The abscissa is the x_n value and the ordinate is the x_{n+1} value. Also in this figure is a line, $x_{n+1} = x_n$. The value for A is 2.4. This is a plot of the *first iterate*, that is, for an initial x_n value, the value for x_{n+1} was found from the equation and the point (x_n, x_{n+1}) was plotted on the graph.

In addition to the curve and line plotted, I have drawn zigzag lines inside the curve. This is a way to show the fixed point. A fixed point is the intersection of the line $f(x) = x$ and the curve. In this figure, the fixed points are at $x = 0$ and $x = 0.5833$. The zigzag lines are used to determine if the fixed points are stable or unstable.

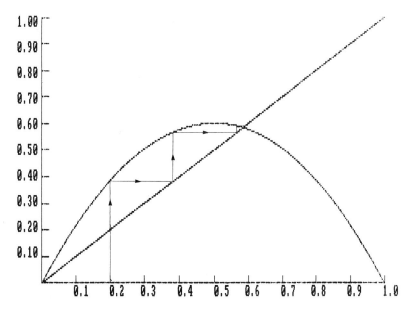

Fig. 2-1. Example of stable point or attractor for the logistic map. Parameter value A = 2.4.

To determine if the points are stable, draw a vertical line from some point on the x-axis until the line intersects the curve. Then draw a horizontal line until you intersect the diagonal line. Repeat these two steps indefinitely. The points in Fig. 2-1 are stable because we have zeroed in on one point, the fixed point. The fixed point in Fig. 2-2 is unstable because we have moved away. You can determine analytically if points will be fixed or stable by an examination of the slope of the curve at the point in question. This can be summarized as follows:

If $\left|\dfrac{df(x)}{dx}\right|_{x=x^*}$ < 1 then x^* is stable.

If $\left|\dfrac{df(x)}{dx}\right|_{x=x^*}$ > 1 then x^* is unstable.

If $\left|\dfrac{df(x)}{dx}\right|_{x=x^*}$ = 1 then x^* is metastable.

If $\left|\dfrac{df(x)}{dx}\right|_{x=x^*}$ = 0 then x^* is superstable.

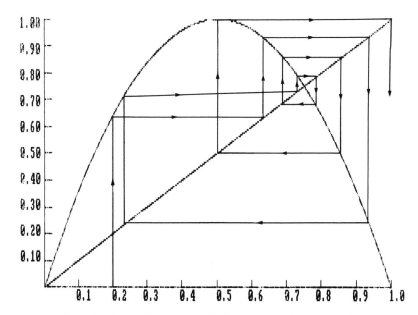

Fig. 2-2.   Example of unstable point for the logistic map. Parameter value A = 4.0.

At the unstable points, the iterates can enter a cycle of period 2, 4, 8, . . . , infinity. The period can double until infinity. There is more on this period doubling in the next section, but for now I would like to discuss the concept of fixed points and cycles and lead into bifurcation theory.

When the iterates enter a cycle of period two, there are two points in the attractor, each said to have equal energy. One iteration leads from one point to the next point. Cycles and fixed points are called attractors. Some attractors are very strange, as you will see in Chapter 3. When the solution to the equation changes at a fixed value, or critical value of a parameter, there is a bifurcation. A point in parameter space where such an event occurs is defined as a *bifurcation point*. At the bifurcation point, two or more solutions emerge for the function.

Let's use a little calculus to study the logistic map in order to determine the critical values for the parameter A and determine the bifurcation points. Given the logistic map

$$f(x) = ax(1 - x)$$

define the first derivative as follows:

$$\lambda = \left| \frac{df(x)}{dx} \right|$$

then

$$\frac{df(x)}{dx} = a - 2ax$$

at $x = 0$, $\quad \lambda = a$

at $x = 1 - \dfrac{1}{a}$, $\quad \lambda = 2 - a$

$a = 3 \quad \lambda = 1 \qquad$ metastable point

$a < 3 \quad \lambda < 1 \qquad$ stable point

$a > 3 \quad \lambda > 1 \qquad$ unstable point

From the above it is clear that at $a = 3$ there is an unstable point. I will now show graphically that a cycle of period two is born at this point.

I have written a small program to create data files that can then be plotted using the program PLOT1, discussed in Chapter 1. This data file when plotted shows the cyclic behavior of the logistic map at various values of the parameter A. The program XVSN creates the data files for the value of the function versus the nth iterate.

XVSN

```
10 CLS
20 INPUT "INPUT FILE NAME ";FILE$
30 INPUT "INPUT PARAMATER A ";A
40 INPUT "INPUT INITIAL VALUE OF X ";X
50 OPEN "O",#1,FILE$
60 FOR I=1 TO 100
70 X=A*X*(1-X)
80 PRINT #1,I,X
90 NEXT I
100 CLOSE #1
110 END
```

The program XVSN first clears the screen in line 10. Then the user enters the name of the data file, a value for the critical parameter, A, and an initial value for x. In line 50 the data file is opened and in line 60 a loop begins. In line 70 the iterate is calculated, and in line 80 the value of the iterate and the iterate number are printed to the file. In line 90 the loop continues. The file is then closed in line 100 and the program ends in line 110.

After a run of the program, the data file can be plotted using the program from Chapter 1, PLOT1. I modified line 470 in that program to read as follows:

470 LINE (X(I − 1),Y(I − 1)) − (X(I),Y(I))

This change connects the data points to give the graph a better visual appearance.

I have made several plots for the first 100 iterates using various values of the parameter A and various initial values for x in the logistic map. These figures were made using the program XVSN and the modified version of PLOT1. Notice that in Fig. 2-3 the function quickly approaches the fixed point of zero. In Fig. 2-4 the fixed point 0.583 is approached, and Fig. 2-5 shows the same value approached using a different initial value. This is an important point. The attractor is approached no matter what the initial value is.

At $A = 3.0$ there is a two-cycle approached, just as I have shown analytically with calculus. This is shown in Fig. 2-6 and 2-7 using two different initial values for x. In Figs. 2-8, 2-9, and 2-10 the same two-cycle has been entered using three different initial values for x, and in Fig. 2-11 a four-cycle has been entered with a critical parameter value of $A = 3.5$.

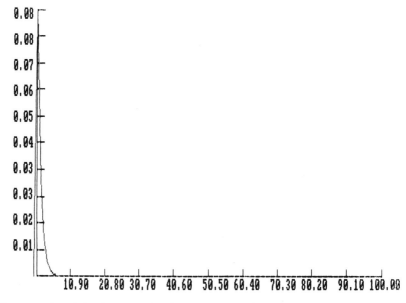

Fig. 2-3.   Plot of the limiting value for parameter value A = 0.4 at an initial value of x = 0.7. The limiting value is 0.

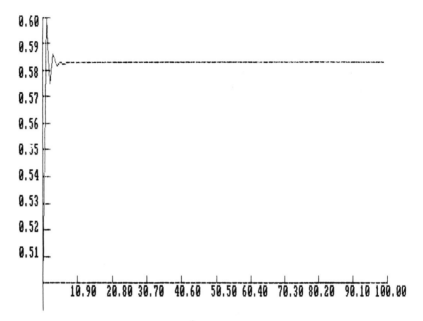

Fig. 2-4. Plot of the limiting value for parameter value A = 2.4 at an initial value of x = 0.7. The limiting value is 0.583.

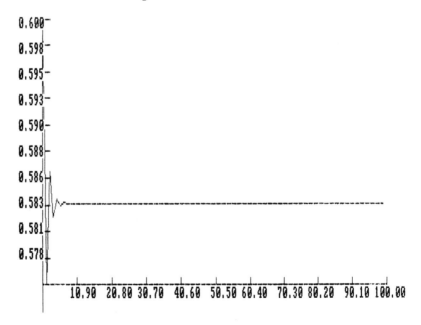

Fig. 2-5. Plot of the limiting value for parameter value A = 2.4 at an initial value of x = 0.5. The limiting value of 0.583 is the same as shown in Fig. 2-4. This shows the effects of a different initial condition.

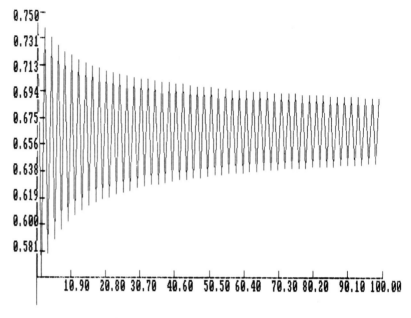

Fig. 2-6.   A two-cycle is approached at the bifurcation point. A = 3, x_o = 0.5.

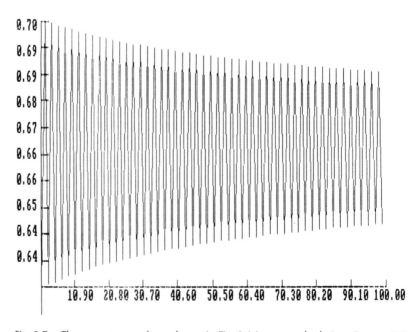

Fig. 2-7.   The same two-cycle as shown in Fig. 2-6 is approached. A = 3, x_o = 0.7.

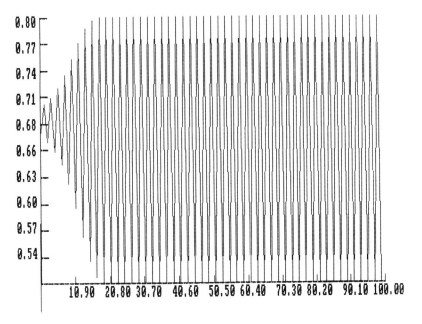

Fig. 2-8. Another two-cycle has been approached. This figure and Figs. 2-9 and 2-10 are the same two-cycle with different starting values for x. A = 3.2, x_o = 0.7.

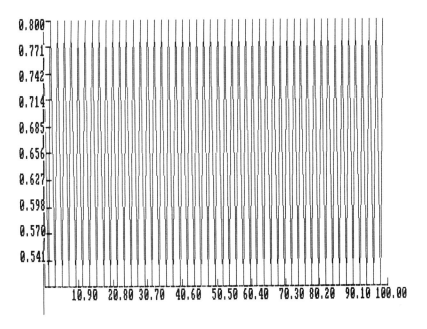

Fig. 2-9. A = 3.2, x_o = 0.5.

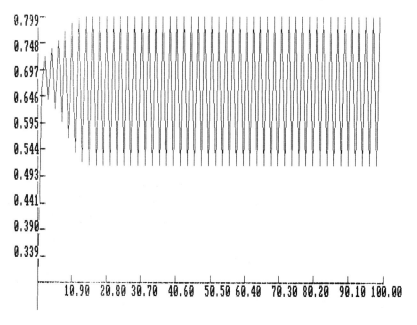

Fig. 2-10.   A = 3.2, x_o = 0.2.

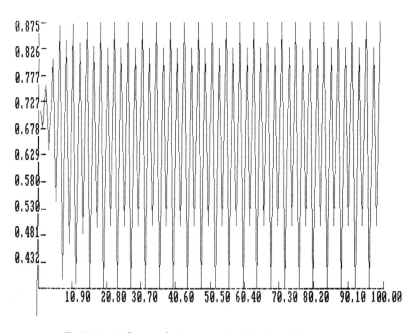

Fig. 2-11.   A four-cycle is approached for A = 3.5, x_o = 0.7.

At parameter values of $A = 3.6$ and $A = 3.8$, the attractor appears to be chaotic, as can be seen in Figs. 2-12, 2-13, and 2-14.

Observation of this behavior suggests a phase diagram or bifurcation diagram for the critical parameter versus the calculated x value. The program PHASE creates a data file that can then be plotted using the program PLOT1.

PHASE

```
10  CLS
15  INPUT "INPUT INITIAL X VALUE ";X
20  INPUT "INPUT FILE NAME ";FILE$
50  OPEN "O",#1,FILE$
52  FOR A=2.8 TO 4! STEP .025
60  FOR I=1 TO 1000
70  X=A*X*(1-X)
72  IF I>980 THEN 80 ELSE 90
80  PRINT #1,A,X
85  N=N+1
90  NEXT I
95  NEXT A
100 CLOSE #1
105 PRINT N
110 END
```

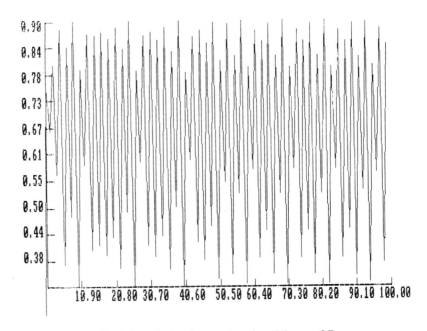

Fig. 2-12.   A chaotic attractor. $A = 3.6$, $x_o = 0.7$.

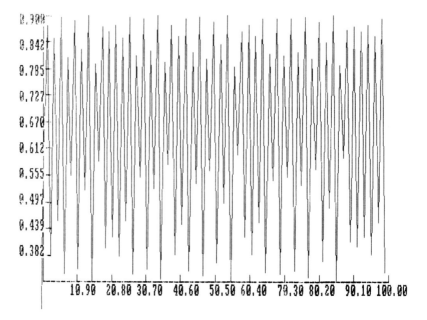

Fig. 2-13.   A chaotic attractor. A = 3.6, x_o = 0.2.

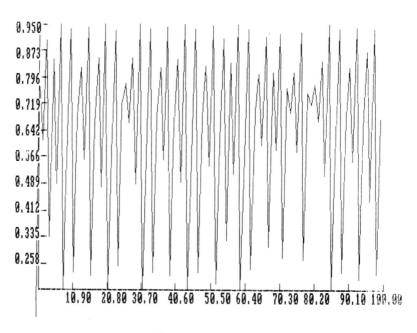

Fig. 2-14.   A chaotic attractor. A = 3.8, x_o = 0.7.

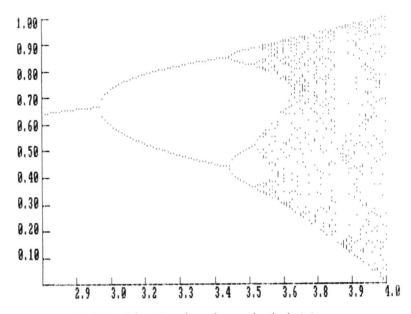

Fig. 2-15.   Bifurcation phase diagram for the logistic map.

In this program, 1000 iterations are calculated for each parameter value *A* from 2.8 to 4.0 in increments of 0.025. The first 980 iterates are disregarded, and the last 20 iterates are written to the data file along with the parameter value. The first 980 iterates are disregarded so that the function will have settled into an attractor, cycle, or chaotic region of phase space. I ran the program PHASE and then used the program PLOT1 to plot the data file created. Fig. 2-15 is the phase diagram for the logistic map.

PERIOD DOUBLING AND SCALING BEHAVIOR

This section answers the question of what the second and higher iterates look like, and examines the universal behavior of nonlinear systems discovered by Feigenbaum (1980).

In order to examine the higher iterates, I have written a computer program to make the plots. The program ITEMAP2 is a general program to generate data files for one-dimensional maps. The program is very simple. After clearing the screen, the program asks the user to input the file name, the critical parameter, and the generation number. The generation number might need some explanation. If you want to examine the first iterate, you enter generation 1. The second iterate is generation 2, and the nth iterate is generation *n*. I use the term "generation" because the logistic map is a model of population dynamics. Other models can easily be substituted for this by changing the appropriate lines in the program.

ITEMAP2

```
10 CLS
20 PRINT "********  DATA FILE GENERATION PROGRAM *********"
30 PRINT "USED TO GENERATE ITERATED MAPS FILES"
40 PRINT
50 PRINT
60 INPUT "INPUT FILE NAME ";FILE$
70 INPUT "INPUT PARAMATER A ";A
80 INPUT "INPUT GENERATION NUMBER (1-20) ";G
90 OPEN "O",#1,FILE$
100 FOR X=0 TO 1 STEP .001
110 '     MAPPING EQUATIONS HERE
120 Y=A*X*(1-X)
125 IF G=1 THEN GOTO 320
130 Y=A*Y*(1-Y)
135 IF G=2 THEN GOTO 320
140 Y=A*Y*(1-Y)
145 IF G=3 THEN GOTO 320
150 Y=A*Y*(1-Y)
155 IF G=4 THEN GOTO 320
160 Y=A*Y*(1-Y)
165 IF G=5 THEN GOTO 320
170 Y=A*Y*(1-Y)
175 IF G=6 THEN GOTO 320
180 Y=A*Y*(1-Y)
185 IF G=7 THEN GOTO 320
190 Y=A*Y*(1-Y)
195 IF G=8 THEN GOTO 320
200 Y=A*Y*(1-Y)
205 IF G=9 THEN GOTO 320
210 Y=A*Y*(1-Y)
215 IF G=10 THEN GOTO 320
220 Y=A*Y*(1-Y)
225 IF G=11 THEN GOTO 320
230 Y=A*Y*(1-Y)
235 IF G=12 THEN GOTO 320
240 Y=A*Y*(1-Y)
245 IF G=13 THEN GOTO 320
250 Y=A*Y*(1-Y)
255 IF G=14 THEN GOTO 320
260 Y=A*Y*(1-Y)
265 IF G=15 THEN GOTO 320
270 Y=A*Y*(1-Y)
275 IF G=16 THEN GOTO 320
```

```
280 Y=A*Y*(1-Y)
285 IF G=17 THEN GOTO 320
290 Y=A*Y*(1-Y)
295 IF G=18 THEN GOTO 320
300 Y=A*Y*(1-Y)
305 IF G=19 THEN GOTO 320
310 Y=A*Y*(1-Y)
315 IF G=20 THEN GOTO 320
320 PRINT #1,X,Y
330 NEXT X
340 CLOSE #1
420 END
```

The data file is opened in line 90 and the calculation begins in a loop in line 100. Data is written to the file in line 320, and the calculation continues until 1000 data points have been calculated. The program ends after closing the file.

Graphs of the first iterate are shown in Figs. 2-16 to 2-21. If you project a line, $f(x) = x$, you can see that there is only one fixed point for each of these graphs. The fixed point may be unstable, depending on the value of the critical parameter. The second iterates are shown in Figs. 2-22 to 2-27. Higher iterates are shown in Figs. 2-28 to 2-39. From these figures and the phase diagram (Fig. 2-15), it is clear that period-doubling occurs

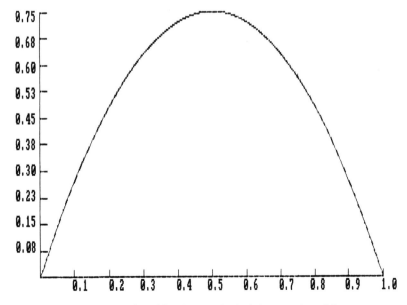

Fig. 2-16. Plot of first iterates for logistic map. A = 3.0.

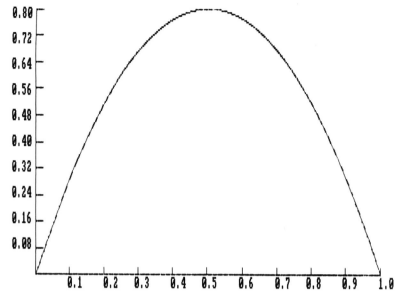

Fig. 2-17. Plot of first iterates for logistic map. A = 3.2.

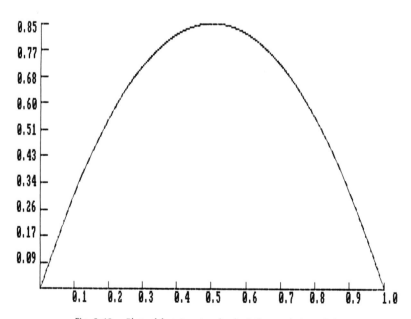

Fig. 2-18. Plot of first iterates for logistic map. A = 3.4.

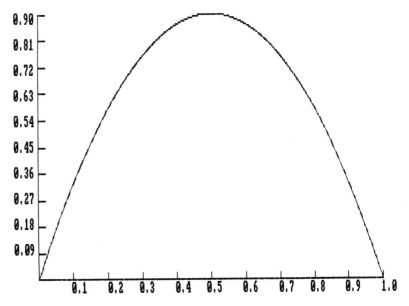

Fig. 2-19. Plot of first iterates for logistic map. A = 3.6.

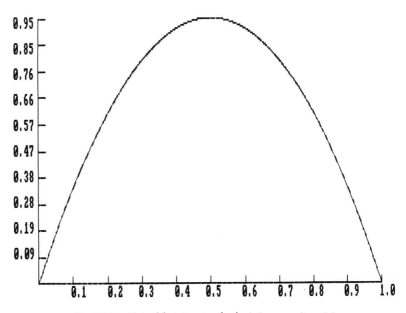

Fig. 2-20. Plot of first iterates for logistic map. A = 3.8.

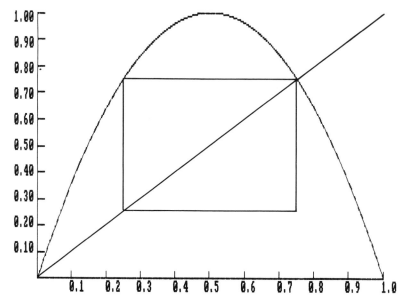

Fig. 2-21.   Plot of first iterates for logistic map. A = 4.0.

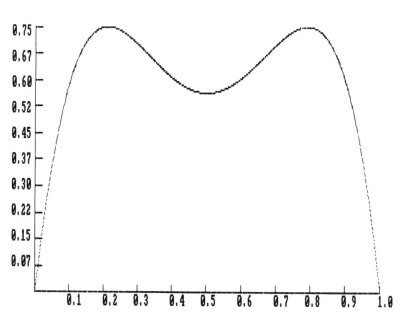

Fig. 2-22.   Plot of second iterates for logistic map. A = 3.0.

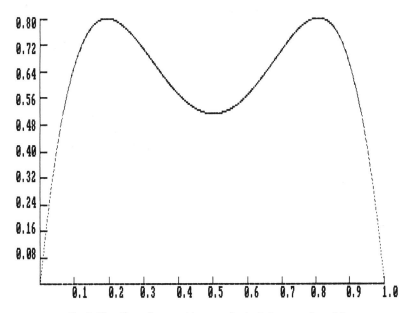

Fig. 2-23.   Plot of second iterates for logistic map. A = 3.2.

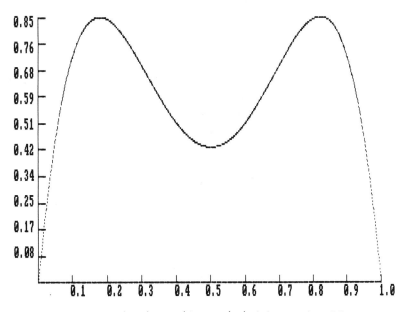

Fig. 2-24.   Plot of second iterates for logistic map. A = 3.4.

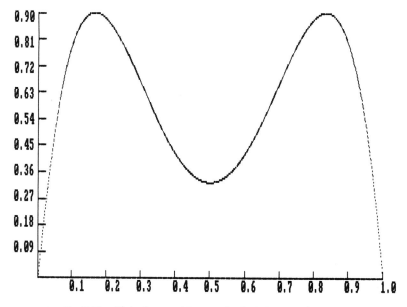

Fig. 2-25.   Plot of second iterates for logistic map. A = 3.6.

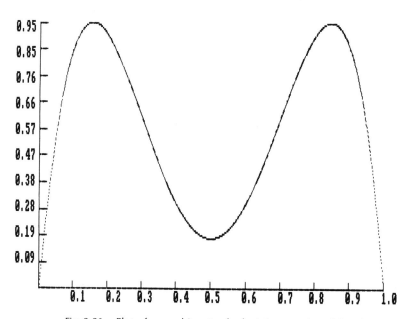

Fig. 2-26.   Plot of second iterates for logistic map. A = 3.8.

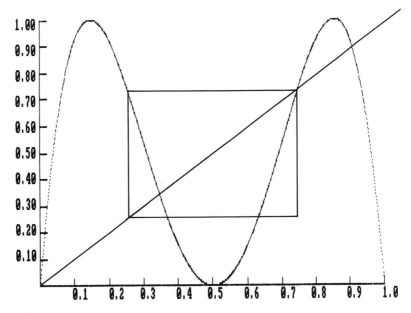

Fig. 2-27.   Plot of second iterates for logistic map. A = 4.0.

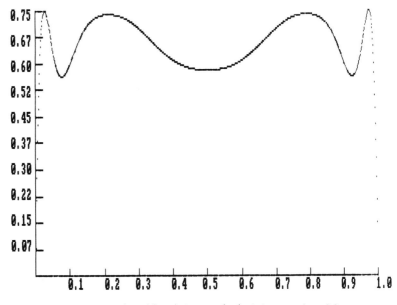

Fig. 2-28.   Plot of fourth iterates for logistic map. A = 3.0.

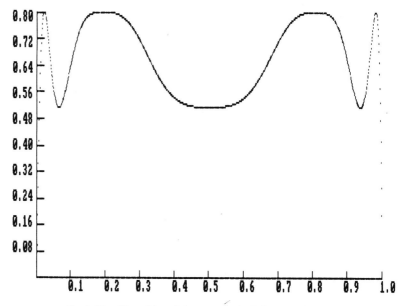

Fig. 2-29. Plot of fourth iterates for logistic map. A = 3.2.

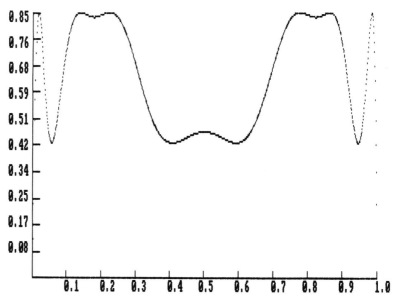

Fig. 2-30. Plot of fourth iterates for logistic map. A = 3.4.

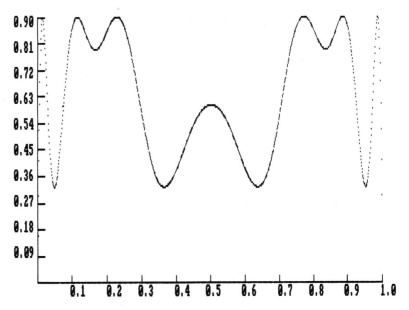

Fig. 2-31.   Plot of fourth iterates for logistic map. A = 3.6.

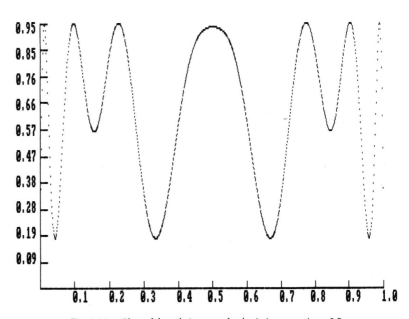

Fig. 2-32.   Plot of fourth iterates for logistic map. A = 3.8.

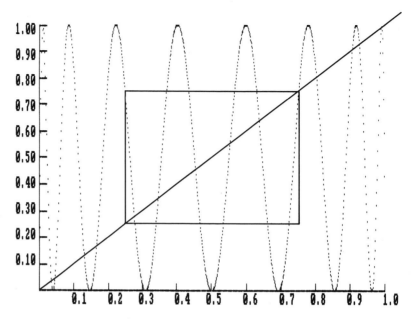

Fig. 2-33.   Plot of fourth iterates for logistic map. A = 4.0.

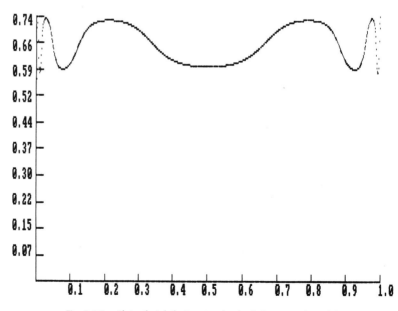

Fig. 2-34.   Plot of eighth iterates for logistic map. A = 3.0.

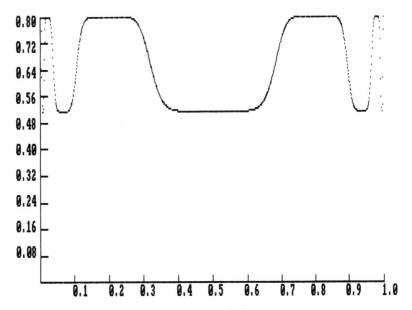

Fig. 2-35.   Plot of eighth iterates for logistic map. A = 3.2.

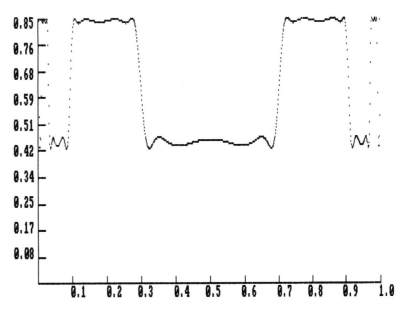

Fig. 2-36.   Plot of eighth iterates for logistic map. A = 3.4.

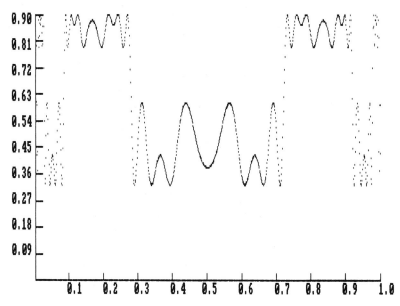

Fig. 2-37.   Plot of eighth iterates for logistic map. A = 3.6.

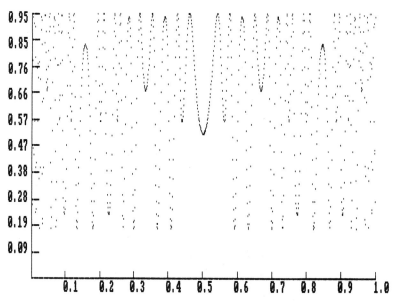

Fig. 2-38.   Plot of eighth iterates for logistic map. A = 3.8.

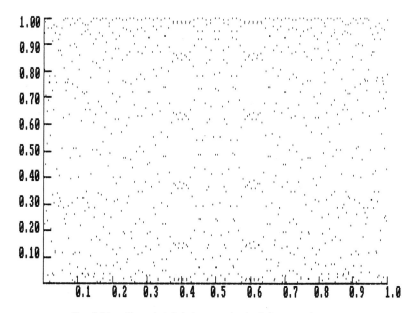

Fig. 2-39.   Plot of eighth iterates for logistic map. A = 4.0.

until chaos is reached. In other words, period-doubling bifurcation is a route to chaos, and the simple logistic map can have quite complicated behavior.

In Figs. 2-40 and 2-41, I superimposed the first and second iterates and drawn the line f(x) = x. In addition, I sketched in a box with a fixed point at one corner. Notice that as the critical parameter increases, the lower portion of the second iterate protrudes from the box. Also notice, in Fig. 2-40, that the portion of the curve inside the box is proportional and of the same shape as the entire first iterate curve.

From all these figures it is clear that as the critical parameter increases, new fixed points are born at the second iterate. These new fixed points will eventually bifurcate, producing a cycle of period four. This can be seen in the fourth iterate plots. Inside the box of the second iterate there is a graph that resembles the first iterate. Similar behavior could be found for the fourth and eighth iterates. Inside small boxes on the fourth and eighth iterates you could find sections of the curve that resemble the entire first iterate curve.

This self-similar behavior, also known as a *fractal*, suggests that a limiting value can be found. What is that limiting value? The entire heuristic argument above suggests a scaling or renormalization rule governing the

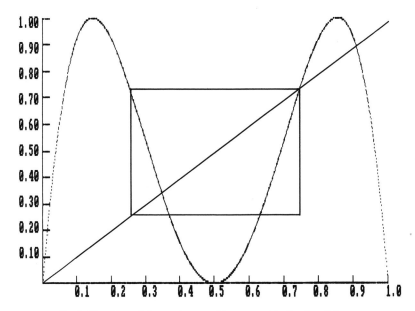

Fig. 2-40.   Plot of first and second iterates with the line f(x) = x.

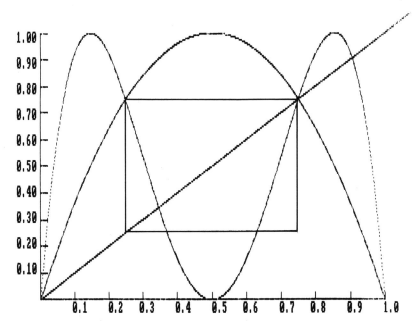

Fig. 2-41.   Plot of first and second iterates with the line f(x) = x. This figure and Fig. 2.40 also show boxes sketched in to illustrate the rescaling behavior of the logistic map.

bifurcation behavior of the logistic map. Let's examine the fixed points closest to x = 0.5. On Fig. 2-42, I have reproduced the phase diagram and sketched in a line corresponding to x = 0.5. The points where this line intersects the bifurcation diagram are fixed points.

Notice that each bifurcation point gives rise to two branches. As the critical parameter increases, the density of bifurcation and fixed points increase. In order to determine the density at high critical parameter values, the equation can be written as follows:

$$d_n = x_n^* - \frac{1}{2}$$

This equation is a measure of the distance of the nearest fixed point from the fixed point x = 0.5. Figure 2-42 includes the first two values for the distance measure, d_n. Feigenbaum defined the ratio of the distance as follows:

$$\alpha = \lim_{n \to \infty} - \frac{d_n}{d_{n+1}}$$

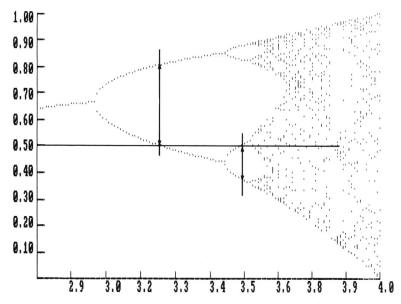

Fig. 2-42. Bifurcation diagram for the logistic map with lines sketched in to allow estimating the Feigenbaum constants. $d_1/d_2 = 2.47$.

Taking the ratio d_1/d_2 gives an approximation to the constant alpha:

$$\alpha \approx \frac{d_1}{d_2} = \frac{42}{17} = 2.47$$

This is in close agreement to Feigenbaum's constant:

$$\alpha = 2.50290787509589 28485 \ldots$$

This is an important constant describing the period-doubling route to chaos. It is in fact a universal constant to describe this type of phenomena. You might want to experiment with other one-dimensional maps such as those listed at the end of this chapter.

Yet another universal constant can be discovered from this bifurcation diagram. Notice that at critical parameter values of $A_1 = 3.00$, $A_2 = 3.448$ and $A_3 = 3.520$, the first, second, and third bifurcations occur. Feigenbaum (1980) defined a constant by the following relation:

$$\delta = \lim_{n \to \infty} \frac{A_{n+1} - A_n}{A_{n+2} - A_{n+1}}$$

For our logistic map you can estimate this universal constant as follows:

$$\delta \approx \frac{3.448 - 3.000}{3.520 - 3.448} = 6.222$$

Feigenbaum calculated this as a limit and obtained the following:

$$\delta = 4.66920160910299097$$

I have plotted a bifurcation diagram for another map, called the quadratic map. This map is one of those listed in the back of this chapter. The map is as follows:

$$f(x) = 1 - Ax^2$$

The bifurcation diagram is shown in Fig. 2-43. I estimated the universal constants as follows:

$$\alpha = \frac{74}{20} = 3.7$$

$$\delta = \frac{1.2 - 0.7}{1.32 - 1.2} = 4.17$$

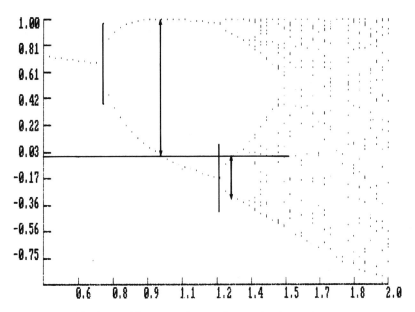

Fig. 2-43.   Bifurcation diagram for another iterated map.

$$\sigma \approx \frac{91}{21} = 4.333, \; \alpha \approx \frac{1.28 - 0.72}{1.38 - 1.28} \approx 5.60$$

In this chapter, I have introduced the one-dimensional maps and derived the universal constants to describe the period-doubling route to chaos.

CATALOG OF ONE-DIMENSIONAL MAPS

The following is a list of several one-dimensional maps that you might want to investigate on your computer. These come from Gould and Tobochnik (1988) and Devaney (1986).

$$f(x) = xe^{r(1-x)}, \quad x_n \in [0,1], \quad 2 < r < 2.7$$

$$f(x) = r[1 - (2x - 1)^4], \quad x_n \in [0,1], \quad 0 < r < 1$$

$$f(x) = re^x$$

$$f(x) = r \sin (x)$$

$$f(x) = rx(1 - x)$$

$$f(x) = rx - x^3$$

$$f(x) = 1 - rx^2$$

3

Strange Attractors

This chapter discusses the concept of attractors and limit cycles in more detail. It shows how to compute the Liapounov exponent and discusses its relevance to characterization of attractors. The concept of fractal dimension and information dimension are also discussed. After all these concepts are covered, several two-and three-dimensional maps and strange attractors are discussed. The concepts and examples are illustrated with computer-generated mappings. The final section of this chapter provides a catalog of strange attractors.

LIAPOUNOV EXPONENTS AND ATTRACTORS

Chapter 2 introduced the idea of attractors. This chapter goes into more detail on attractor points, limit cycles, and strange attractors. Chapter 2 also introduced the idea of using the derivative of the function at fixed points to determine the stability of the system at that point. This idea is developed in more detail here with the introduction of Liapounov exponents. A.M. Liapounov was a Russian mathematician who lived between 1857 and 1918. (Many authors spell his name as "Lyapunov.")

In Chapter 2, Fig. 2-38—a mapping of an interval to itself—shows that at a critical parameter value of $A = 3.8$, the eighth iterate mapping is chaotic and the second iterate for this parameter value is a two-cycle. The characterization of the attractor can be deduced from the derivative of the slope, as shown in Chapter 2, or from the Liapounov exponent. This is defined by Lauwerier (1986) as follows:

$$\sigma = \lim_{n \to \infty} \frac{1}{n} \sum_{k=0}^{n-1} \log \left| \frac{df}{dx} \right|$$

or

$$\sigma = \int \log \left| \frac{df}{dx} \right| d\mu$$

For the logistic map, the Liapounov exponent is calculated as

$$x_{n+1} = 4x_n(1 - x_n)$$

$$\sigma = \frac{1}{\pi} \int_0^1 \frac{\log \left| 4(1 - 2x) \right|}{\sqrt{4(1 - x)}}$$

$$= \int_0^1 \log \left(4 \cos \left(\frac{\pi \delta}{2} \right) \right) d\delta = \log 2$$

where δ is a dummy of integration.

The Liapounov exponent is a measure of the exponential separation between two adjacent initial points. This is also a measure of the loss of information after one iteration. I will clarify these concepts later in this section. Now I would like to reintroduce and delve more deeply into the concept of the attractor and limit cycle, and show how the Liapounov exponent characterizes the type of attractor.

Figure 2-2 shows an example of an unstable point, and Fig. 2-1 is an example of a stable point. Both are from the one-dimensional logistic map. A two-dimensional iterated mapping can also have a stable and unstable point. These are sketched in Fig. 3-1.

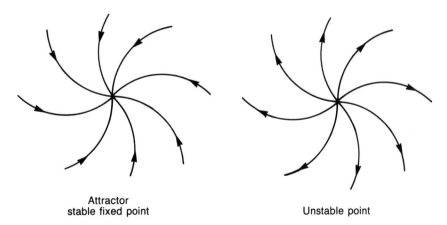

Attractor
stable fixed point

Unstable point

Fig. 3-1. Examples of fixed points.

For the attractor or stable point, all the trajectories fall into the attractor point. Thus the name, attractor. The unstable points cause all trajectories to escape from the point.

All nearby initial points are attracted to the fixed points and repelled by the unstable point. This could be thought of as an energy surface where the stable attractor is a valley or pit and the unstable point is a peak or ridge in the energy surface. A particle on such an energy surface will come to rest at the attractor point. If the particle has enough energy and there are two or more attractor points of equal depth, the particle may oscillate from one fixed point to another. These are best not called fixed points because the particle does not rest at any one of them. This local repulsion and global attraction of the particle flow implies the formation of a closed curve around the unstable point. This is shown schematically in Fig. 3-2.

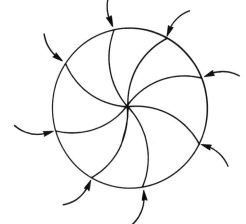

Fig. 3-2.   Schematic of limit cycle.

The limit cycle shown in this figure is a circle, but this does not imply that all limit cycles are circles. Other shapes, some very odd, are possible. The important point to note about limit cycles is that the resulting attractor is a periodic cycle. Figure 2-6 is a plot of data that represents a simple two cycle. This can be thought of as a limit cycle. The two- cycle bifurcates into a four cycle as illustrated in Fig. 2-11.

Often in nonlinear dynamics, the terminology is some what different. Figure 3-3 shows four attractors, also called *nodes*.

There are at least two other types of attractors. One is called a *T2 torus*, shown in Figure 3-4, and the other is called a strange attractor, many of which are shown in this chapter.

A system with friction or an analogy of friction is called a dissipative system. Frictionless systems are called *conservative* or *Hamiltonian systems*.

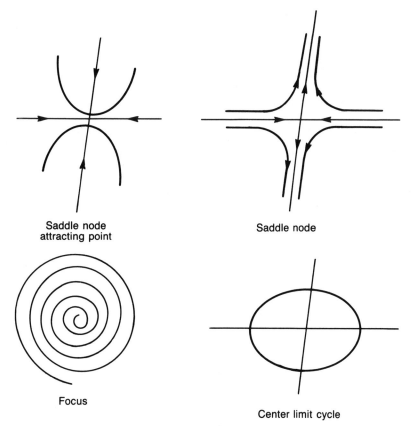

Fig. 3-3.   Examples of invariant curves.

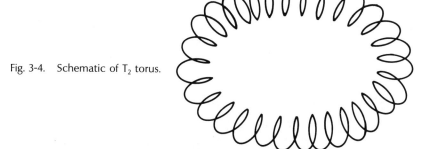

Fig. 3-4.   Schematic of T_2 torus.

In dissipative systems, the phase-space volumes are contracted in time. This phase-space volume contraction does not imply equal contraction in all directions. Some directions may contract while others may stretch, such that the final volume is less than the initial volume. This also implies that in a dissipative system the final motion may be unstable within the attractor. This contraction in volume—stretching in some directions and contraction in others—usually results in an exponential separation of orbits of points that are initially very close on the attractor. The exponential separation takes place in the direction of stretching.

This exponential separation results in what is called *sensitive dependence to initial conditions*. This is the primary characteristic of strange attractors. The system will appear to be chaotic. Sensitivity to initial conditions is also the accepted definition of strange attractors. A computer plot of a strange attractor appears to be chaotic. This is now the accepted definition of chaos.

Earlier in this chapter, I introduced the Liapounov exponent with the one-dimensional logistic map as an example. I would now like to reintroduce this concept with respect to two or more dimensions. Suppose you are given the following two-dimensional map

$$x_{n+1} = f(x_n, y_n)$$

$$y_{n+1} = g(x_n, y_n)$$

The local behavior at a fixed point can be determined from the Jacobian:

$$J = \begin{bmatrix} \partial f/\partial x & \partial f/\partial y \\ \partial g/\partial x & \partial g/\partial y \end{bmatrix}$$

The eigenvalues λ_1, λ_2, etc. of the Jacobian matrix are called the *multipliers* and the log of the eigenvalues are called the *Liapounov exponents*. Sensitive dependence on initial conditions corresponds to at least one characteristic exponent being greater than zero.

Let's look at these ideas in more detail. If λ_1 and λ_2 are the eigenvalues of the Jacobian matrix J, and the Jacobian determinate is constant, then the following is true:

$$\lambda_1 + \lambda_2 = \log |J|$$

Furthermore, the rate of contraction of the volume element in phase space is the rate of change of the Jacobian determinate J.

There are as many characteristic exponents as there are dimensions in the phase space of the dynamical system. As stated earlier in this chapter, the characteristic exponents measure the exponential behavior of the trajectories in phase space. The magnitude of an attractor's characteristic exponents is a measure of its degree of chaos. If the characteristic exponent is negative or positive, it determines the measure of convergence or divergence of the trajectories on the attractor. This is summed up in Table 3-1 for a three-dimensional phase space dynamical system.

Table 3-1

Sign of Exponent			Type of Attractor	Dimension of Attractor
−	−	−	Fixed point	Zero
0	−	−	Limit cycle	One
0	0	−	Two torus	Two
+	0	−	Strange attractor	Two

Throughout the rest of this chapter I will make use of some or all of the ideas in Table 3-1 while discussing specific mappings of dynamical systems.

THE DIMENSION OF STRANGE ATTRACTORS

As a first step in determining the dimension of an attractor, Froehling, et al. (1981) suggests that the number of nonnegative characteristic exponents represents the dimension of an attractor. These dimensions are listed in Table 3-1 for the fixed point, the limit cycle, the two torus, and the strange attractor. The strange attractor actually has a folded structure and a fractal dimension greater than two and less than three.

Let's see how to calculate the fractal dimension, also known as *Hausdorf dimension*, of a strange attractor. As a first example of determining dimension, take the example of a cube. Take a cube and double its linear size in each direction. The volume is now eight times greater:

$$2^3 = 8$$

In general,

$$k = l^D$$

where an object of dimension D has its sides increased in each spatial dimension by a factor of l and the volume is given by k. The dimension is now given by the following:

$$D = \frac{\log k}{\log l}$$

This implies that D need not be an integer. It is thus possible to have a fractional or fractal dimension.

As another example, take the Cantor set illustrated in Fig. 1-1. You can calculate the dimension of its limiting set of points. Increasing the linear size by a factor of $l = 3$ will give $k = 2$, two copies of the same object:

$$D_H = \frac{\log 2}{\log 3} \approx 0.625$$

Let's look at another example, the classical fractal curve known as the Koch curve, shown in Fig. 3-5. The Koch curve has an infinite perimeter, but is bounded to a region of the plane. In this case the dimension is given by

$$D_H = \frac{\log 4}{\log 3} \approx 1.26$$

Fig. 3-5.   Koch curve—a simple fractal.

To calculate the dimension numerically, you divide the phase space into cells of linear size, e, then count the number, N, of cells that contain at least one point of the orbit. The dimension is then given by the following:

$$D_H = \lim_{e \to 0} \frac{\log (N(e))}{\log \left(\frac{1}{e}\right)}$$

This relationship can be found from the slope of a plot of log N versus log $1/e$. You count the N for a series of e's and make the appropriate plot on a log-log scale.

For an attractor with a fractal dimension D_i, the information gained in a measurement with a resolution e is given by the following

$$I = D_i |\log (e)|$$

This leads to a definition of the information dimension:

$$D_I = \lim_{e \to 0} \frac{I(e)}{|\log (e)|}$$

The concept of information dimension and the sensitivity dependence on initial conditions leads to the observation that order and information can be created by strange attractors. You could speculate that planets and life forms were created by strange attractors formed by the Big Bang.

LIMIT CYCLES: EXAMPLES

Dimensions of chaotic attractors are discussed at length by Farmer (1982). The rest of this chapter expands on many of the ideas covered in the previous two sections, in connection with specific examples of dynamic systems.

Rayleigh System

This section examines computer-generated plots of three limit cycles. The first example is the Rayleigh system given by the following equations:

$$\frac{dx}{dt} = y$$

$$\frac{dy}{dt} = \left(y - \frac{y^3}{3}\right) - x + f \cos (\theta)$$

$$\frac{d\theta}{dt} = w$$

This system, like all limit cycles, is an example of self-organization. If you start the initial condition anywhere in the phase space, the system quickly settles to the limit cycle like that shown in Fig. 3-6. This plot was made with the program RAYLEIGH. This program is a combination of two programs previously introduced and discussed in Chapter 1. The program

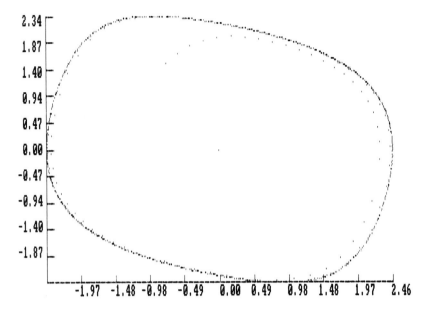

Fig. 3-6. Rayleigh system limit cycle.

SDEQ1 and PLOT1 have been combined, with the file storage and reading removed.

RAYLEIGH

```
10  CLS
20  DIM X(2001),Y(2001),Z(2001)
30  X=-1:Y=+1
40  FOR T9=0 TO 100 STEP .1
50  PRINT INT(T9*10),X,Y,Z
60  FOR T=T9 TO T9+.1 STEP .1/25
70  D1=Y
80  D2=(Y-(Y*Y*Y)/3)-X+1*COS(Z)
90  D3=1
100 X=X+D1*.1/25
110 Y=Y+D2*.1/25
120 Z=Z+D3*.1/25
130 X(INT(T9*10))=X
140 Y(INT(T9*10))=Y
150 Z(INT(T9*10))=Z
160 NEXT T
170 NEXT T9
180 XMAX=-1E+20 :XMIN=-XMAX
190 YMAX=-1E+20
200 YMIN=-YMAX
210 NPTS=1000
```

```
220 FOR I=1 TO NPTS
230 IF YMIN>Y(I) THEN YMIN=Y(I)
240 IF YMAX<Y(I) THEN YMAX=Y(I)
250 IF XMAX<X(I) THEN XMAX=X(I)
260 IF XMIN>X(I) THEN XMIN=X(I)
270 NEXT I
280 CLS
290 NXTIC=10:NYTIC=10
300 XMN=XMIN:XMX=XMAX:YMN=YMIN:YMX=YMAX
310 CLS
320 SCREEN 2:KEY OFF
330 DSX=ABS(XMX-XMN):DSY=ABS(YMX-YMN)
340 SX=.1:SY=.1
350 AXMN=XMN-DSX*SX:AXMX=XMX+DSX*SX
360 AYMX=YMX+DSY*SY:AYMN=YMN-DSY*SY
370 WINDOW (AXMN,AYMN)-(AXMX,AYMX)
380 LINE (XMN,YMN)-(XMX,YMN)
390 LINE (XMN,YMN)-(XMN,YMX)
400 DXTIC=DSX*.02:DYTIC=DSY*.025
410 XTIC=DSX/NXTIC:YTIC=DSY/NYTIC
420 FOR I=1 TO NXTIC
430 XP=XMN+XTIC*I
440 LINE (XP,YMN)-(XP,YMN+DYTIC)
450 ROW=24
460 NEXT I
470 FOR I= 1 TO NYTIC
480 YP=YMN+I*YTIC
490 LINE (XMN,YP)-(XMN+DXTIC,YP)
500 NEXT I
510 FOR I=1 TO NPTS-1
520 J=I+1
530 IF Y(I)>YMX OR Y(J)>YMX OR Y(J)<YMN OR X(I)<XMN
    THEN 550
540 CIRCLE (X(I),Y(I)),0
550 NEXT I
560 FOR I=1 TO NXTIC
570 XP=XMN+XTIC*I
580 XC=PMAP(XP,0)
590 COL=INT(80*XC/640)-1
600 LOCATE 24,COL
610 PRINT USING "###.#"; XP;
620 NEXT I
630 FOR I=1 TO NYTIC
640 YP=YMN+I*YTIC
650 YR=PMAP(YP,1)
660 ROW=CINT(24*YR/199)+1
670 LOCATE ROW,1
680 PRINT USING "###.#"; YP
690 NEXT I
700 GOTO 700
```

Van der Pol Oscillator

The second limit cycle example is the Van der Pol system. This system was first described by Van der Pol (1926, 1927) to describe the buildup of oscillations in a nonlinear electrical circuit. Hayashi (1985) has described this system at length, and Batten (1987) has described an analog computer model of the Van der Pol system. This system is described by the following equations:

$$\frac{dx}{dt} = y$$

$$\frac{dy}{dt} = \left(-\frac{1}{cl}\right)(x + (3Bx^2 - A)y) + f\sin(\theta)$$

$$\frac{d\theta}{dt} = w$$

As you have seen for the Rayleigh system, this limit cycle program, VANDERPO, is a combination of SDEQ1 and PLOT1 with file handling removed. The plot of the Van der Pol limit cycle is shown in Fig. 3-7.

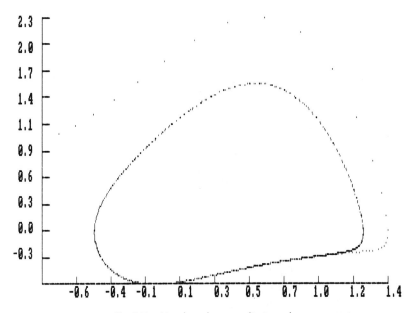

Fig. 3-7. Vanderpol system limit cycle.

VANDERPO

```
10 CLS
20 DIM X(2001),Y(2001),Z(2001)
30 X=-1:Y=+1
40 FOR T9=0 TO 100 STEP .1
50 PRINT INT(T9*10),X,Y,Z
60 FOR T=T9 TO T9+.1 STEP .1/25
70 D1=Y
80 D2=-1*(X+(3*X*X-1)*Y)+.55*SIN(1)
90 D3=1
100 X=X+D1*.1/25
110 Y=Y+D2*.1/25
120 Z=Z+D3*.1/25
130 X(INT(T9*10))=X
140 Y(INT(T9*10))=Y
150 Z(INT(T9*10))=Z
160 NEXT T
170 NEXT T9
180 XMAX=-1E+20 :XMIN=-XMAX
190 YMAX=-1E+20
200 YMIN=-YMAX
210 NPTS=1000
220 FOR I=1 TO NPTS
230 IF YMIN>Y(I) THEN YMIN=Y(I)
240 IF YMAX<Y(I) THEN YMAX=Y(I)
250 IF XMAX<X(I) THEN XMAX=X(I)
260 IF XMIN>X(I) THEN XMIN=X(I)
270 NEXT I
280 CLS
290 NXTIC=10:NYTIC=10
300 XMN=XMIN:XMX=XMAX:YMN=YMIN:YMX=YMAX
310 CLS
320 SCREEN 2:KEY OFF
330 DSX=ABS(XMX-XMN):DSY=ABS(YMX-YMN)
340 SX=.1:SY=.1
350 AXMN=XMN-DSX*SX:AXMX=XMX+DSX*SX
360 AYMX=YMX+DSY*SY:AYMN=YMN-DSY*SY
370 WINDOW (AXMN,AYMN)-(AXMX,AYMX)
380 LINE (XMN,YMN)-(XMX,YMN)
390 LINE (XMN,YMN)-(XMN,YMX)
400 DXTIC=DSX*.02:DYTIC=DSY*.025
410 XTIC=DSX/NXTIC:YTIC=DSY/NYTIC
420 FOR I=1 TO NXTIC
430 XP=XMN+XTIC*I
440 LINE (XP,YMN)-(XP,YMN+DYTIC)
450 ROW=24
460 NEXT I
```

```
470 FOR I= 1 TO NYTIC
480 YP=YMN+I*YTIC
490 LINE (XMN,YP)-(XMN+DXTIC,YP)
500 NEXT I
510 FOR I=1 TO NPTS-1
520 J=I+1
530 IF Y(I)>YMX OR Y(J)>YMX OR Y(J)<YMN OR X(I)<XMN
    THEN 550
540 CIRCLE (X(I),Y(I)),0
550 NEXT I
560 FOR I=1 TO NXTIC
570 XP=XMN+XTIC*I
580 XC=PMAP(XP,0)
590 COL=INT(80*XC/640)-1
600 LOCATE 24,COL
610 PRINT USING "###.#"; XP;
620 NEXT I
630 FOR I=1 TO NYTIC
640 YP=YMN+I*YTIC
650 YR=PMAP(YP,1)
660 ROW=CINT(24*YR/199)+1
670 LOCATE ROW,1
680 PRINT USING "###.#"; YP
690 NEXT I
700 GOTO 700
```

Brusselator

As a third and final example of limit cycles, I will describe the Brusselator. The Brusselator was given its name as a tribute to Prigogine and Lefever. (Prigogine won the 1977 Nobel prize in chemistry for his work in nonlinear systems and thermodynamics.) Danby (1985) uses the Brusselator system as one of his examples of nonlinear differential equations. The system has been described at length by Babloyantz (1986) in *Molecules, Dynamics, and Life: An Introduction to Self-Organization of Matter*. The Brusselator system is a model for a chemical reaction given by the following chemical schematic

$$A \rightarrow X + B$$

$$X \rightarrow Y + C$$

$$2X + Y \rightarrow 3X + D$$

$$X \rightarrow E + F$$

From these equations we can write the differential equations for the concentration of the chemical species:

$$\frac{dx}{dt} = [A] \, ([B] + 1)X + X^2Y$$

$$\frac{dy}{dt} = [B]X - X^2Y$$

In the program BRUSS I have selected the following values for the constants:

$$A = 1$$

$$B = 3$$

This produces the plot shown in Fig. 3-8. This is a limit cycle. You might want to run some other computer experiments by changing the values for the parameters A and B in this system and observing the results.

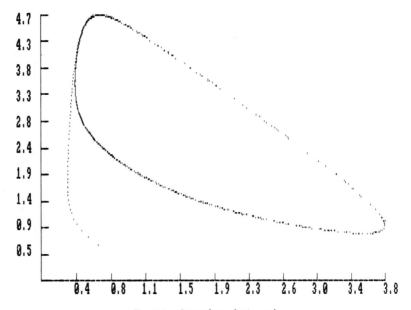

Fig. 3-8.   Brusselator limit cycle.

BRUSS

```
10 CLS
20 DIM X(2001),Y(2001)
30 X=1:Y=.2
40 FOR T9=0 TO 100 STEP .1
50 PRINT INT(T9*10),X,Y
60 FOR T=T9 TO T9+.1 STEP .1/25
70 D1=1-(3+1)*X+X*X*Y
80 D2=3*X-X*X*Y
100 X=X+D1*.1/25
110 Y=Y+D2*.1/25
130 X(INT(T9*10))=X
140 Y(INT(T9*10))=Y
160 NEXT T
170 NEXT T9
180 XMAX=-1E+20 :XMIN=-XMAX
190 YMAX=-1E+20
200 YMIN=-YMAX
210 NPTS=1000
220 FOR I=1 TO NPTS
230 IF YMIN>Y(I) THEN YMIN=Y(I)
240 IF YMAX<Y(I) THEN YMAX=Y(I)
250 IF XMAX<X(I) THEN XMAX=X(I)
260 IF XMIN>X(I) THEN XMIN=X(I)
270 NEXT I
280 CLS
290 NXTIC=10:NYTIC=10
300 XMN=XMIN:XMX=XMAX:YMN=YMIN:YMX=YMAX
310 CLS
320 SCREEN 2:KEY OFF
330 DSX=ABS(XMX-XMN):DSY=ABS(YMX-YMN)
340 SX=.1:SY=.1
350 AXMN=XMN-DSX*SX:AXMX=XMX+DSX*SX
360 AYMX=YMX+DSY*SY:AYMN=YMN-DSY*SY
370 WINDOW (AXMN,AYMN)-(AXMX,AYMX)
380 LINE (XMN,YMN)-(XMX,YMN)
390 LINE (XMN,YMN)-(XMN,YMX)
400 DXTIC=DSX*.02:DYTIC=DSY*.025
410 XTIC=DSX/NXTIC:YTIC=DSY/NYTIC
420 FOR I=1 TO NXTIC
430 XP=XMN+XTIC*I
440 LINE (XP,YMN)-(XP,YMN+DYTIC)
450 ROW=24
460 NEXT I
470 FOR I= 1 TO NYTIC
480 YP=YMN+I*YTIC
490 LINE (XMN,YP)-(XMN+DXTIC,YP)
500 NEXT I
510 FOR I=1 TO NPTS-1
520 J=I+1
```

```
530 IF Y(I)>YMX OR Y(J)>YMX OR Y(J)<YMN OR X(I)<XMN
    THEN 550
540 CIRCLE (X(I),Y(I)),0
550 NEXT I
560 FOR I=1 TO NXTIC
570 XP=XMN+XTIC*I
580 XC=PMAP(XP,0)
590 COL=INT(80*XC/640)-1
600 LOCATE 24,COL
610 PRINT USING "###.#"; XP;
620 NEXT I
630 FOR I=1 TO NYTIC
640 YP=YMN+I*YTIC
650 YR=PMAP(YP,1)
660 ROW=CINT(24*YR/199)+1
670 LOCATE ROW,1
680 PRINT USING "###.#"; YP
690 NEXT I
700 GOTO 700
```

This, like all limit cycles, is a self-organizing system. You might want to experiment with other initial conditions in this system and the limit cycles to observe the insensitivity of the initial conditions on the final limit cycle in the phase space. Recall that the system is self-organizing.

STRANGE ATTRACTORS: EXAMPLES

In this section, several strange attractors are discussed in various degrees of detail. A strange attractor, above all else, contains a sensitivity to initial conditions. Two initial points can quickly diverge into apparent chaos; for this reason, this is sometimes called *deterministic chaos*. Another property of strange attractors is that an infinite number of iterations will not escape from the phase space, but rather wander within the confined phase volume. This is due to the folding and stretching of phase space, as was discussed earlier in this chapter.

Some strange attractors are fractal and exhibit self-similarity under magnification, and some are not fractals. All strange attractors have at least one positive Liapounov exponent.

Duffing Oscillator

The first strange attractor modeled here is the Duffing oscillator. This is a forced oscillator with a cubic term:

$$\frac{dx}{dt} = y$$

$$\frac{dy}{dt} = -(ax^3 + cx + bx) + f \cos(\theta)$$

I use the program DUFFING to study the behavior of this system. The program, like others introduced in this chapter, is made from combining SDEQ1 with PLOT1 and removing the sections concerning file reading and writing.

DUFFING

```
10 CLS
20 DIM X(2001),Y(2001),Z(2001)
30 X=-1:Y=+1
40 FOR T9=0 TO 200 STEP .1
50 PRINT INT(T9*10),X,Y,Z
60 FOR T=T9 TO T9+.1 STEP .1/25
70 D1=Y
80 D2=-(-.05*X*X*X+X+.2*Y)+1.6*COS(Z)
90 D3=1
100 X=X+D1*.1/25
110 Y=Y+D2*.1/25
120 Z=Z+D3*.1/25
130 X(INT(T9*10))=X
140 Y(INT(T9*10))=Y
150 Z(INT(T9*10))=Z
160 NEXT T
170 NEXT T9
180 XMAX=-1E+20 :XMIN=-XMAX
190 YMAX=-1E+20
200 YMIN=-YMAX
210 NPTS=2000
220 FOR I=1 TO NPTS
230 IF YMIN>Y(I) THEN YMIN=Y(I)
240 IF YMAX<Y(I) THEN YMAX=Y(I)
250 IF XMAX<X(I) THEN XMAX=X(I)
260 IF XMIN>X(I) THEN XMIN=X(I)
270 NEXT I
280 CLS
290 NXTIC=10:NYTIC=10
300 XMN=XMIN:XMX=XMAX:YMN=YMIN:YMX=YMAX
310 CLS
320 SCREEN 2:KEY OFF
330 DSX=ABS(XMX-XMN):DSY=ABS(YMX-YMN)
340 SX=.1:SY=.1
350 AXMN=XMN-DSX*SX:AXMX=XMX+DSX*SX
360 AYMX=YMX+DSY*SY:AYMN=YMN-DSY*SY
370 WINDOW (AXMN,AYMN)-(AXMX,AYMX)
380 LINE (XMN,YMN)-(XMX,YMN)
390 LINE (XMN,YMN)-(XMN,YMX)
400 DXTIC=DSX*.02:DYTIC=DSY*.025
```

```
410 XTIC=DSX/NXTIC:YTIC=DSY/NYTIC
420 FOR I=1 TO NXTIC
430 XP=XMN+XTIC*I
440 LINE (XP,YMN)-(XP,YMN+DYTIC)
450 ROW=24
460 NEXT I
470 FOR I= 1 TO NYTIC
480 YP=YMN+I*YTIC
490 LINE (XMN,YP)-(XMN+DXTIC,YP)
500 NEXT I
510 FOR I=1 TO NPTS-1
520 J=I+1
530 IF Y(I)>YMX OR Y(J)>YMX OR Y(J)<YMN OR X(I)<XMN
    THEN 550
540 CIRCLE (X(I),Y(I)),0
550 NEXT I
560 FOR I=1 TO NXTIC
570 XP=XMN+XTIC*I
580 XC=PMAP(XP,0)
590 COL=INT(80*XC/640)-1
600 LOCATE 24,COL
610 PRINT USING "###.##"; XP;
620 NEXT I
630 FOR I=1 TO NYTIC
640 YP=YMN+I*YTIC
650 YR=PMAP(YP,1)
660 ROW=CINT(24*YR/199)+1
670 LOCATE ROW,1
680 PRINT USING "###.##"; YP
690 NEXT I
700 GOTO 700
```

Figure 3-9 will suggests some of the rich behavior of a forced oscillator. The figure shows chaotic orbits and a periodic orbit, as shown by the apparent limit cycle. Figure 3-10 is a strange attractor of the Duffing system with the following parameter values:

$$a = 1.0$$

$$b = 0.3$$

$$c = 0$$

$$F = 10.0$$

These values are entered by changing the appropriate values in the equations in program line 80. You might want to experiment with other values

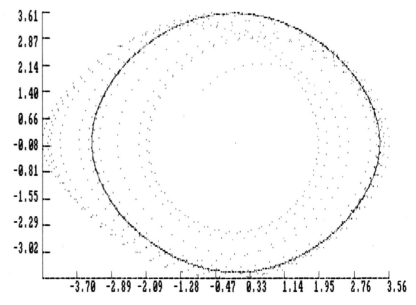

Fig. 3-9.   Duffing oscillator. F = 1.6, a = −0.05, b = 0.2, c = 1.0.

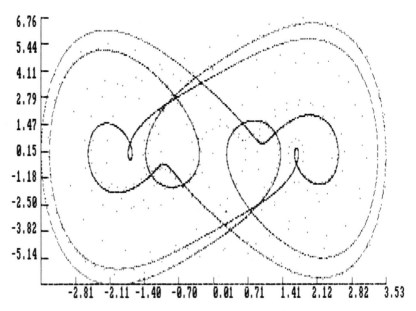

Fig. 3-10.   Duffing oscillator. F = 10.0, a = 1.0, b = 0.3, c = 0.

for these parameters to observe other strange attractors and limit cycles for this system.

Henon Attractor

The next strange attractor discussed here is the Henon attractor. This attractor was introduced by Henon (1976) and has been studied by Curry (1979), Ruelle (1980), Thompson and Thompson (1980), and Devaney (1986).

The Henon map is a two-dimensional analog of the quadratic map introduced in Chapter 1. The mapping is given by the following relation:

$$x_{n+1} = y_n + 1 - ax_n^2$$

$$y_{n+1} = bx_n$$

The system is shown in Fig. 3-11. It is an area-contracting mapping of a strange attractor. Iterating to infinity will not cause the points to diverge to infinity; rather they will always wander on the chaotic attractor.

The program I used to make Fig. 3-11 is called HENON1. This is a modified version of ITEMAP2, discussed in Chapter 2. Line 80 opens a file

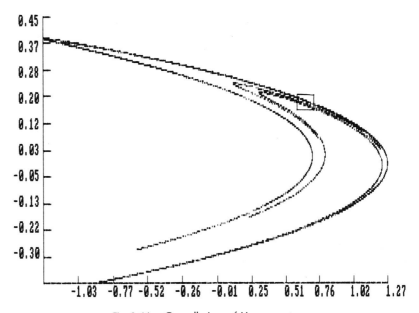

Fig. 3-11. Overall view of Henon system.

of the name chosen in line 70. Lines 90 and 100 set the initial conditions, and a loop for 1000 iterations begins in line 100. In line 280 the data points are printed to a file. The loop continues in line 310 and the file closes in line 420.

HENON1

```
10 CLS
20 DIM X(1001),Y(1001)
30 PRINT "********  DATA FILE GENERATION PROGRAM *********"
40 PRINT "USED TO GENERATE ITERATED MAPS FILES"
70 INPUT "INPUT FILE NAME ";FILE$
80 OPEN "O",#1,FILE$
90 X(1)=1.5
100 Y(1)=1.5
110 FOR T=1 TO 1000
130 X(T+1)=Y(T)+1-1.4*X(T)*X(T)
140 Y(T+1)=.3*X(T)
150 X(I)=X(T+1)
160 Y(I)=Y(T+1)
280 PRINT #1,X(I),Y(I)
290 I=I+1
300 PRINT T,X(I),Y(I)
310 NEXT T
320 CLOSE #1
400 END
```

Let's look at the Henon map in more detail. Earlier in this chapter I pointed out that the local behavior at a fixed point can be determined from the Jacobian:

$$J = \begin{bmatrix} \partial f/\partial x & \partial f/\partial y \\ \partial g/\partial x & \partial g/\partial y \end{bmatrix}$$

For the Henon mapping; the following is true:

$$T\begin{pmatrix} x_{n+1} \\ y_{n+1} \end{pmatrix} = \begin{pmatrix} 1 - ax_n^2 + y_n \\ bx_n \end{pmatrix}$$

The Jacobian, therefore, is given by

$$J = \begin{bmatrix} -2ax & 1 \\ b & 0 \end{bmatrix} = -b$$

If $|b| < 1$, the area is contracted by a factor $|b|$ at each iteration. Since $|b| = 0.3$, the contraction is not great, and we can see the fractal structure of the Henon attractor. Other strange attractors have such strong contraction (10^{-6}) that fractal structure, if it is present, can not be seen. The mapping has two invariant points, given by the following:

$$x_{n+1} = x_n$$

$$y_{n+1} = y_n$$

or

$$x = \frac{-(1 - b) \pm \sqrt{(1 - b)^2 + 4a}}{2a}$$

$$y = bx$$

These two points are real for

$$a > a_o = \frac{(1 - b)^2}{4}$$

Under this condition, one of the points is linearly unstable and the other is unstable for

$$a > a_1 = \frac{3(1 - b)^2}{4}$$

Henon determined the parameters from numerical experiments and obtained the following values.

$$a = 1.4$$

$$b = 0.3$$

The Henon mapping is given in Fig. 3-11. The initial conditions are completely irrelevent. For the control parameters $a = 1.4$ and $b = 0.3$, the mapping will always have the same general form. Furthermore, this strange attractor is a fractal. Its fractal dimension is 1.26.

The fractal property of self-similarity under magnification is clearly seen in Figs. 3-11 to 3-13. The first figure, Fig. 3-11, shows the entire Henon

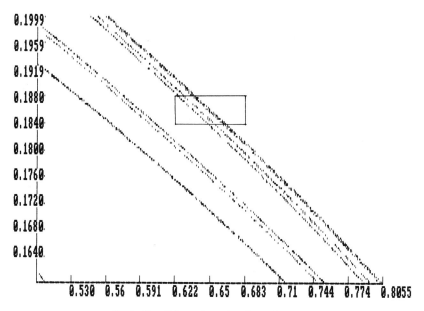

Fig. 3-12. Enlarged section from Fig. 3-11.

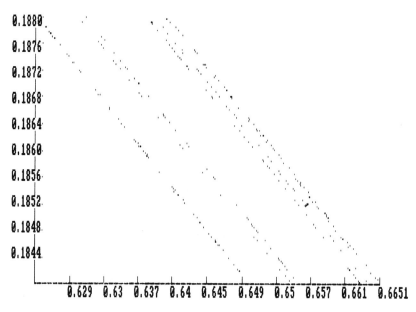

Fig. 3-13. Enlarged section from Fig 3-12.

strange attractor and a small window I have sketched on the mapping. This box represents the window for the magnified plot in Fig. 3-12. On this figure I have again sketched a small box and magnified this window in Fig. 3-13. These plots were made with the program HENON9, which is a simple modification of HENON1. This program includes selecting only those data points within the window to be written to the data file. These figures clearly show the self-similarity under magnification and therefore the fractal nature of the Henon strange attractor.

HENON9

```
10  CLS
20  DIM X(5001),Y(5001)
30    INPUT "INPUT FILE NAME ";FILE$
40    OPEN "O",#1,FILE$
50  U1=1.5
60  V1=1.5
70  FOR I=1 TO 2000
80    U1=U2
90    V1=V2
100 U2=V1+1-1.4*U1*U1
110 V2=.3*U1
120 IF U2<.5 THEN 80
130 IF U2>.9 THEN 80
140 IF V2<.16 THEN 80
150 IF V2>.2 THEN 80
160 X(I)=U2
170 Y(I)=V2
180    PRINT #1,X(I),Y(I)
190 'PRINT I,X(I),Y(I)
200 NEXT I
210    CLOSE #1
220 END
```

Rossler System

The next example of a strange attractor is the Rossler system. Rossler (1976) set out to make a very simple model of a truncated Navier-Stokes equation, similar to that introduced by Tedeschini-Lalli (1982) and given by the following relations:

$$x_1 = -2x_1 + 4\sqrt{5}x_2x_3 + 4\sqrt{5}x_4x_5$$

$$x_2 = -9x_2 + 3\sqrt{5}x_1x_3$$

$$x_3 = -5x_3 - 7\sqrt{5}x_1x_2 + 9Ex_1x_7 + R$$

$$x_4 = -5x_4 - \sqrt{5}x_1x_5$$

$$x_5 = -x_5 - 3\sqrt{5}x_1x_5$$

$$x_6 = -x_6 = 5Ex_1x_5$$

$$x_7 = -5x_7 - 9Ex_1x_3$$

Rossler was inspired by Lorenz (1963). Lorenz also inspired Franceschine and Tebaldi (1979), who devised a five-mode truncation of the Navier-Stokes equations. The Rossler system is given by the following relation:

$$\frac{dx}{dt} = -y - z$$

$$\frac{dy}{dt} = x + ay$$

$$\frac{dz}{dt} = b + xz - cz$$

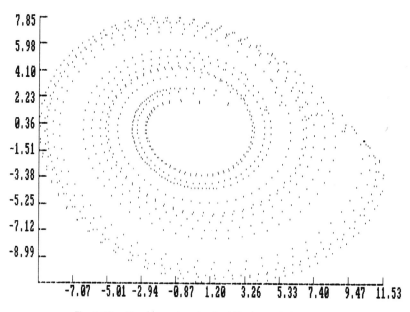

Fig. 3-14.   Rossler system in the XY plane. 1000 points.

A plot of this system is shown in Fig 3-14. The parameters for this plot are as follows:

$$a = 0.2$$

$$b = 0.2$$

$$c = 5.7$$

The plot was made with the program PLOT1 after creating a data file with the program ROSSLER.

ROSSLER

```
10 REM DEFINE DX/DT=D1=F(T,X,Y,Z) IN LINE 140
15 REM DEFINE DY/DT=D2=F(T,X,Y,Z) IN LINE 150
17 REM DEFINE DZ/DT=D3=F(T,X,Y,Z) IN LINE 155
20 INPUT "INPUT INITIAL AND FINAL VALUES OF T ";T1,T2
45 INPUT "INPUT DELTA T ";D
50 INPUT "INPUT INITIAL CONDITIONS X,Y,Z ";X,Y,Z
80 INPUT "INPUT NUMBER OF SILENT CALCULATIONS FOR EACH
DELTA T ";N
90 INPUT "INPUT FILE NAME ";FILE$
95 OPEN "O",#1,FILE$
110 FOR T9=T1 TO T2 STEP D
120 PRINT T9,X,Y,Z
125 PRINT #1,X,Y,Z
130 FOR T=T9 TO T9+D STEP D/N
140 D1=-Y-Z
150 D2=X+Y/5
155 D3=1/5+Z*(X-5.7)
160 X=X+D1*D/N
170 Y=Y+D2*D/N
175 Z=Z+D3*D/N
180 NEXT T
190 NEXT T9
195 CLOSE #1
200 END
```

The program ROSSLER is a modified version of SDEQ1 with parameters a, b, and c fixed. Figure 3-14 contains 1000 data points. Figure 3-15 is a more detailed plot for the same parameters with 3000 data points. Besides having 3000 data points, this plot was made with finer time increments to the solution of the differential equations.

Figures 3-16 and 3-17 are plots of the same data set looking at a different portion of phase space. Figure 3-16 is the YZ plane and Fig.

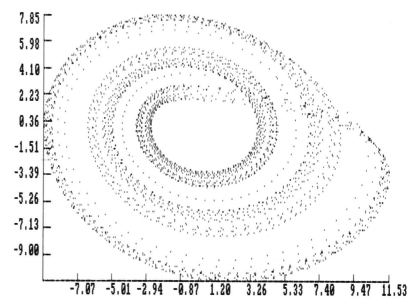

Fig. 3-15. Rossler system in the XY plane. 3000 points.

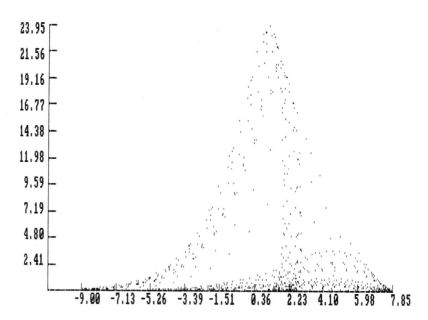

Fig. 3-16. Rossler system in the YZ plane. 3000 points.

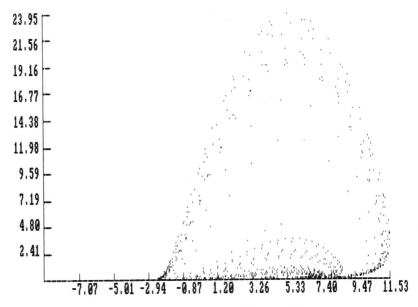

Fig. 3-17.   Rossler system in the XZ plane. 3000 points.

3-17 is the XZ plane. These plots clearly show the three-dimensional structure of this strange attractor. You might want to experiment with the
parameters *a, b,* and *c* in order to observe the sensitivity to parameter
changes that you would expect for a strange attractor.

Henon Mapping

In this strange attractor we will look at the phase-locking route to
chaos. This mapping is also known as a Henon map. Henon (1969) was
interested in the study of the dynamics of clusters of galaxies. He started
with a Cremona transform of the form

$$x_1 = f(x,y)$$

$$y_1 = g(x,y)$$

where the functions *f* and *g* are polynomials. He then made the simplifying
assumption and wrote the mapping transform as follows:

$$x_{n+1} = x_n \cos(A) - (y_n - x_n^2) \sin(A)$$

$$y_{n+1} = y_n \sin(A) + (y_n - x_n^2) \cos(A)$$

The Henon mapping has been discussed by Hughes (1986).

Before I discuss the program for this mapping and some diagrams of the map I would like to mention a number of points including the KAM theorem and frequency-locking. This mapping is a conservative system. This means it is not a dissipative system and is a frictionless system. Such frictionless systems are area-preserving mappings known as *Hamiltonian* systems. These systems have been reviewed in *Chaos* by Bai-Lin (1984).

The motion of a classical conservative system of N degrees of freedom is given by a Hamiltonian function of the form

$$H = H(p_1, \ldots, p_N; q_1, \ldots, q_N)$$

where p_i is the position and q_i is the momentum of particle i. If we transform the set $\{p_i, q_i\}$ into a new set of variables $\{J_i, Q_i\}$ such that in terms of these new variables the Hamiltonian function depends only on J_i's, then all the Q_i's become cyclic variables and the following is true:

$$H = H(J_1, \ldots, J_N)$$

We then have the following equations:

$$\frac{\partial H}{\partial J_i} = \Omega_i(J_1, \ldots, J_N)$$

$$\frac{\partial H}{\partial Q_i} = 0$$

These can be integrated to give the following:

$$Q_i(t) = \Omega_i(t) + Q_i(0)$$

$$J_i(t) = J_i(0)$$

This is said to be an *integrable system*.

Now if the system is non-integrable, the Hamiltonian becomes

$$H = H_o + V(p_i; q_i)$$

where H_o is integrable and V is a small parameter that has dependence on the variables, known as a perturbation. These systems have been studied by Kalmogorov (1954), Arnold (1963), and Moser (1962). They evolved the KAM theorem. They assumed that the perturbation V is small and the frequencies Ω_i of the unperturbed system satisfy the following nonresonance condition:

$$\frac{\partial(\Omega_1, \ldots \Omega_N)}{\partial(J_1, \ldots, J_N)} = 0$$

The motion is then confined to an N-torus. These N-tori are called the KAM surfaces. For systems of greater than two degrees of freedom, the trajectories may wander along the whole energy surface, generating a chaos known as *Arnold diffusion*.

If we take two coupled nonlinear oscillators in the nearly integrable region, a section of the plane space would correspond to motion of the two different frequencies called *resonance zones*. With strong coupling, the resonance zones tend to overlap and chaotic layers of finite width are formed. The KAM torus then looks something like the schematic shown in Fig. 3-18.

Fig. 3-18. Example of a KAM curve.

In short, the KAM theorem states that under small perturbations, a Hamiltonian system will remain stable except for small bands of instability that correspond to resonance between the original system and the disturbance. The resonances occur when the ratio of the periods of the two frequencies is a rational number.

For example, asteroids in the asteroid belt are perturbed by the gravitational pull of Jupiter. If two orbits of Jupiter take as long as five orbits of an asteroid, then there is a 2/5 resonance. An asteroid caught between resonances may go into a chaotic orbit or be thrown out of orbit and escape to become a comet. The gaps in the asteroid belt, known as the *Kirkwood gaps*, are believed to be caused by this phenomena.

As I pointed out earlier, when the phase-locking occurs, the ratio between the frequencies is a rational number. A plot of the frequency of the oscillator against the frequency of the perturbing force gives rise to a staircase-type plot known as the *Devil's staircase*. Bak (1986) has discussed this in some-detail with examples from solid state physics. Figure 3-19 shows the sketch of a schematic of the Devil's staircase. As you can see from the figure, there is a fractal structure as shown by the self-similarity under magnification.

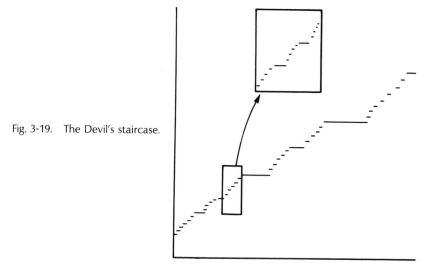

Fig. 3-19. The Devil's staircase.

Now let's look at some cross sections of KAM tori and see the periodic, phase-locked, and chaotic orbits that an orbiting object (a particle in a magnetic field, an asteroid, a galaxy, etc.) may take. The program HENON2 is made from the program PLOT1. The file-reading capability of PLOT1 has been removed and the appropriate code for generating the data to fill two arrays has been added.

HENON2

```
5 CLS
10   DIM X(4001),Y(4000),U(251),V(251)
20 INPUT "INPUT PARAMATER A";A
24 CLS
26 PRINT "CALCULATIONS IN PROGRESS"
30 FOR ORBIT=.1 TO 1.6 STEP .1
50 U(1)=ORBIT/3
60 V(1)=ORBIT/3
70 FOR T=1 TO 250
80 U(T+1)=U(T)*COS(A)-(V(T)-U(T)*U(T))*SIN(A)
90 V(T+1)=U(T)*SIN(A)+(V(T)-U(T)*U(T))*COS(A)
100 Q=Q+1
105 X(Q)=U(T)
106 Y(Q)=V(T)
110 NEXT T
120 NEXT ORBIT
130 XMAX=-1E+20 :XMIN=-XMAX
140 YMAX=-1E+20
150 YMIN=-YMAX
160 NPTS=2500
170 FOR I=1 TO NPTS
180 IF YMIN>Y(I) THEN YMIN=Y(I)
190 IF YMAX<Y(I) THEN YMAX=Y(I)
200 IF XMAX<X(I) THEN XMAX=X(I)
210 IF XMIN>X(I) THEN XMIN=X(I)
220 NEXT I
230 CLS
240 NXTIC=10:NYTIC=10
250 XMN=XMIN:XMX=XMAX:YMN=YMIN:YMX=YMAX
260 CLS
270 SCREEN 2:KEY OFF
280 DSX=ABS(XMX-XMN):DSY=ABS(YMX-YMN)
290 SX=.1:SY=.1
300 AXMN=XMN-DSX*SX:AXMX=XMX+DSX*SX
310 AYMX=YMX+DSY*SY:AYMN=YMN-DSY*SY
320 WINDOW (AXMN,AYMN)-(AXMX,AYMX)
330 LINE (XMN,YMN)-(XMX,YMN)
340 LINE (XMN,YMN)-(XMN,YMX)
350 DXTIC=DSX*.02:DYTIC=DSY*.025
360 XTIC=DSX/NXTIC:YTIC=DSY/NYTIC
370 FOR I=1 TO NXTIC
380 XP=XMN+XTIC*I
390 LINE (XP,YMN)-(XP,YMN+DYTIC)
400 ROW=24
410 NEXT I
420 FOR I= 1 TO NYTIC
```

```
430 YP=YMN+I*YTIC
440 LINE (XMN,YP)-(XMN+DXTIC,YP)
450 NEXT I
460 FOR I=1 TO NPTS-1
470 J=I+1
480 IF Y(I)>YMX OR Y(J)>YMX OR Y(J)<YMN OR X(I)<XMN
THEN 500
490 CIRCLE (X(I),Y(I)),0
500 NEXT I
510 FOR I=1 TO NXTIC
520 XP=XMN+XTIC*I
530 XC=PMAP(XP,0)
540 COL=INT(80*XC/640)-1
550 LOCATE 24,COL
560 PRINT USING "###.##"; XP;
570 NEXT I
580 FOR I=1 TO NYTIC
590 YP=YMN+I*YTIC
600 YR=PMAP(YP,1)
610 ROW=CINT(24*YR/199)+1
620 LOCATE ROW,1
630 PRINT USING "###.##"; YP
640 NEXT I
650 GOTO 650
```

The program starts with a screen clear followed by a DIM statement, and then asks the user to input parameter A. This is the angle in radians between zero and pi. The control parameter called ORBIT, introduced in line 30, has been selected to correspond with the initial conditions in lines 50 and 60. The initial condition is ORBIT/3. The variable ORBIT starts at 0.1 and ends at 1.6 in increments of 0.1.

The equations in line 80 and 90 are iterated 250 times to give a total of 250 points per orbit. These data points are stored in an xy array in lines 105 and 106. Variable Q, in line 100, is a counting variable to fill the arrays. The graphics then begin in line 130, and the rest of the program is similar to PLOT1. By changing the loop parameter ORBIT in line 30, you can effect changes in the initial conditions, and by changing the step parameter in line 30, you can change the number of orbits.

Figure 3-20 is a simple Henon map in which the parameter A is 1.1; A is 1.2 in Fig. 3-21. Notice how the outer orbit is beginning to break up. Extending these results in Fig. 3-22, with A = 1.264, there is an inner orbit that has broken up and the outer orbits are not continuous. When A = 1.265, in Fig. 3-23, the inner orbits are coalescing into islands of stability. These islands are even more clearly seen when A = 1.275 in Fig. 3-24. By the time A = 1.300 in Fig. 3-25, the islands appear to be moving

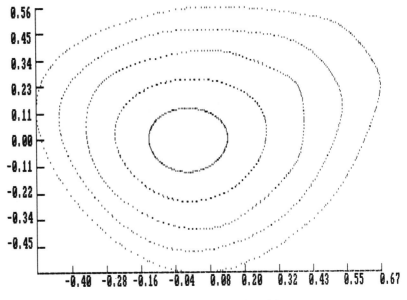

Fig. 3-20.   KAM torus. A = 1.1.

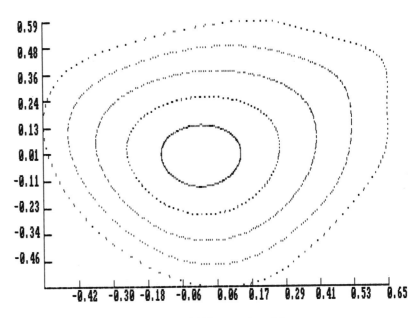

Fig. 3-21.   KAM torus. A = 1.2.

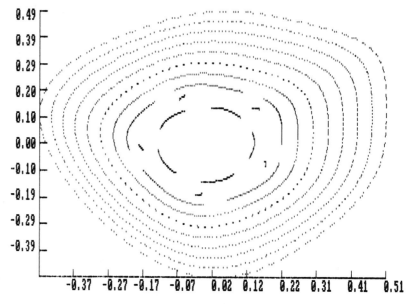

Fig. 3-22.   KAM torus. A = 1.264.

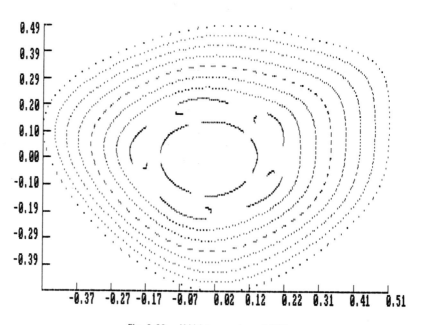

Fig. 3-23.   KAM torus. A = 1.265.

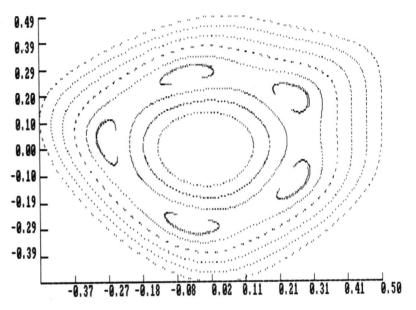

Fig. 3-24.   KAM torus. A = 1.275.

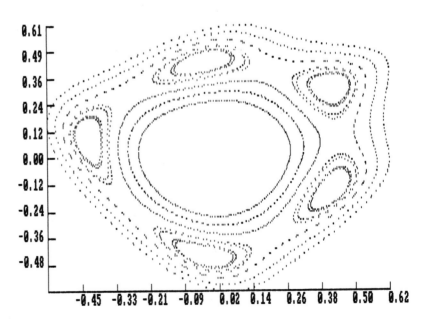

Fig. 3-25.   KAM torus. A = 1.3.

out toward the outer orbit and are perturbing the structure of the outer orbits.

When $A = 1.35$ in Fig. 3-26 the islands have moved out past the outer orbits and are beginning to break up. At $A = 1.370$ (Fig. 3-27) the outer orbits appear to show the effects of the islands breaking up and there appears to be a breakup in an inner orbit. When $A = 1.40$ (Fig. 3-28) the breakup of the inner orbit is more apparent, and the outer orbits show less perturbation. In Fig. 3-29, when $A = 1.50$, the middle orbit appears to be breaking up. At $A = 1.570$ (Fig. 3-3) the breakup is affecting all the orbits and distorting their shape. When $A = 1.575$ the entire system seems to be cooperating to generate chaotic orbits. At $A = 1.580$ (Fig. 3-32) a stable inner orbit has formed and the outer orbits are coalescing to form four chaotic branches. When $A = 1.590$ in Fig. 3-33, there are stable inner orbits and the outer orbits are chaotic. At $A = 1.600$ (Fig. 3-34) the inner orbits seem to be perturbing the chaotic outer orbits and creating order. When $A = 1.601$ (Fig. 3-35) the effects of this perturbation is beginning to be clear. The chaotic outer orbits are breaking up completely and are spinning out in Fig. 3-36, at $A = 1.602$. When $A = 1.700$ (Fig. 3-37) stability seems to have been reached and only the outer orbit is slightly broken.

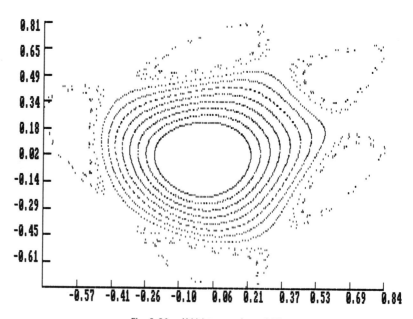

Fig. 3-26.   KAM torus. $A = 1.35$.

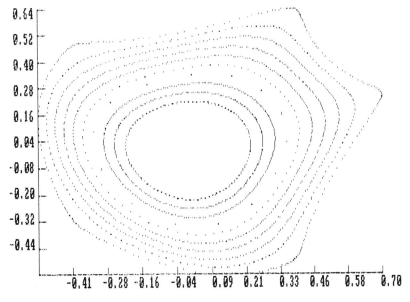

Fig. 3-27.   KAM torus. A = 1.37.

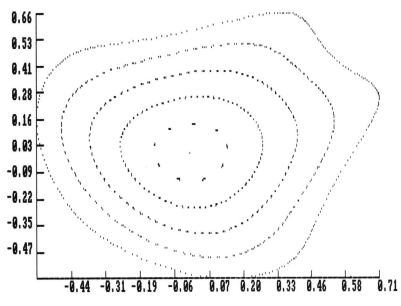

Fig. 3-28.   KAM torus. A = 1.4.

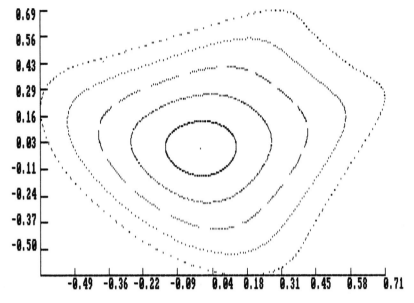

Fig. 3-29.   KAM torus. A = 1.5.

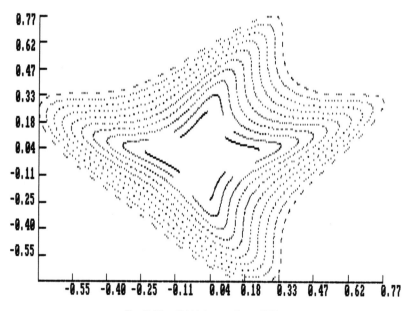

Fig. 3-30.   KAM torus. A = 1.57.

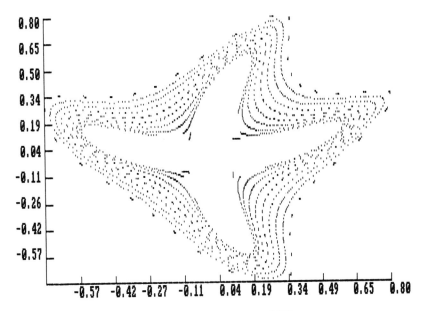

Fig. 3-31.   KAM torus. A = 1.575.

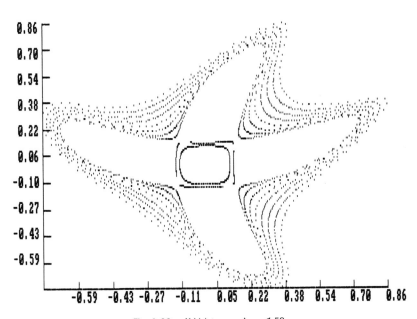

Fig. 3-32.   KAM torus. A = 1.58.

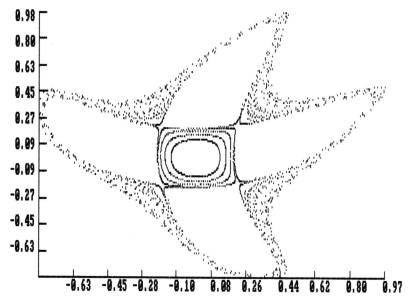

Fig. 3-33.   KAM torus. A = 1.59.

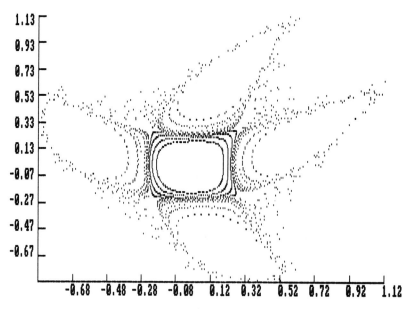

Fig. 3-34.   KAM torus. A = 1.6.

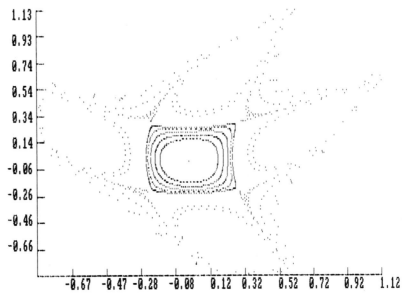

Fig. 3-35.   KAM torus. A = 1.601.

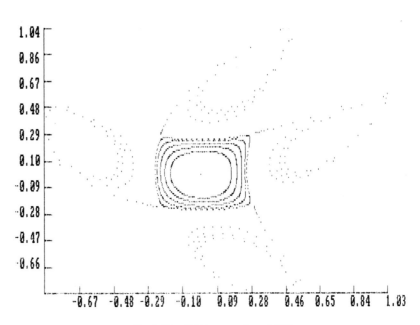

Fig. 3-36.   KAM torus. A = 1.602.

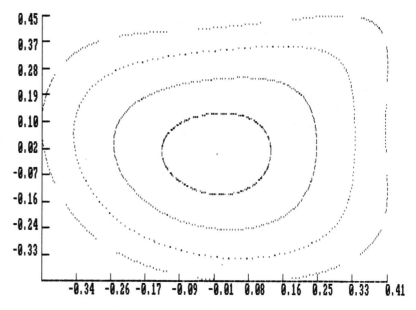

Fig. 3-37.   KAM torus. A = 1.7.

You might want to experiment with changing the initial conditions by changing the variable ORBIT in line 30 and changing the parameter *A*. You could easily explore many regions of these KAM tori.

Lorenz Attractor

The Lorenz attractor is a classic strange attractor. It is one of the first strange attractors reported in 1962 by Lorenz (1963). Lorenz was trying to model atmospheric dynamics of the planet. This model involved a truncated model of the Navier-Stokes equations. The model is as follows:

$$\frac{dx}{dt} = \sigma(y - x)$$

$$\frac{dy}{dt} = rx - y - xz$$

$$\frac{dz}{dt} = xy - bz$$

The Lorenz attractor has been very deeply studied by many people, including Robbins (1979) and Sparrow (1982).

Calculating the divergence of the Lorenz attractor produces the following:

$$\frac{\partial x}{\partial x} = -\sigma$$

$$\frac{\partial y}{\partial y} = -x - 1 \text{ at } x = 0; \quad \frac{\partial y}{\partial y} = -1$$

$$\frac{\partial z}{\partial z} = -b$$

Since the divergence is defined as

$$\nabla \equiv \frac{\partial x}{\partial x} + \frac{\partial y}{\partial y} + \frac{\partial z}{\partial z}$$

we get

$$\frac{\partial x}{\partial x} + \frac{\partial y}{\partial y} + \frac{\partial z}{\partial z} = -(\sigma + b + 1)$$

Since ∇ is negative, the volume in phase space shrinks to zero. In fact, this strange attractor contracts by a multiplier of about 10^{-6} at each iteration. The contraction is so fast that there is no hope of viewing fractal structure even if it exists for this attractor.

Now let's examine the program to plot the strange attractor of Lorenz. The program LORENZ is a combination of SDEQ1 and PLOT1. It is similar to other programs introduced in this chapter; only the equations in lines 70, 80, and 90 have been changes.

LORENZ

```
10 CLS
20 DIM X(5001),Y(5001),Z(5001)
30 X=5:Y=5:Z=5
40 FOR T9=0 TO 500 STEP .1
50 PRINT INT(T9*10),X,Y,Z
60 FOR T=T9 TO T9+.1 STEP .1/25
70 D1=10*(Y-X)
80 D2=28*X-Y-X*Z
90 D3=X*Y-(8/3)*Z
100 X=X+D1*.1/25
110 Y=Y+D2*.1/25
120 Z=Z+D3*.1/25
```

```
130 X(INT(T9*10))=X
140 Y(INT(T9*10))=Z
150 Z(INT(T9*10))=Y
160 NEXT T
170 NEXT T9
180 XMAX=-1E+20 :XMIN=-XMAX
190 YMAX=-1E+20
200 YMIN=-YMAX
210 NPTS=5000
220 FOR I=1 TO NPTS
230 IF YMIN>Y(I) THEN YMIN=Y(I)
240 IF YMAX<Y(I) THEN YMAX=Y(I)
250 IF XMAX<X(I) THEN XMAX=X(I)
260 IF XMIN>X(I) THEN XMIN=X(I)
270 NEXT I
280 CLS
290 NXTIC=10:NYTIC=10
300 XMN=XMIN:XMX=XMAX:YMN=YMIN:YMX=YMAX
310 CLS
320 SCREEN 2:KEY OFF
330 DSX=ABS(XMX-XMN):DSY=ABS(YMX-YMN)
340 SX=.1:SY=.1
350 AXMN=XMN-DSX*SX:AXMX=XMX+DSX*SX
360 AYMX=YMX+DSY*SY:AYMN=YMN-DSY*SY
370 WINDOW (AXMN,AYMN)-(AXMX,AYMX)
380 LINE (XMN,YMN)-(XMX,YMN)
390 LINE (XMN,YMN)-(XMN,YMX)
400 DXTIC=DSX*.02:DYTIC=DSY*.025
410 XTIC=DSX/NXTIC:YTIC=DSY/NYTIC
420 FOR I=1 TO NXTIC
430 XP=XMN+XTIC*I
440 LINE (XP,YMN)-(XP,YMN+DYTIC)
450 ROW=24
460 NEXT I
470 FOR I= 1 TO NYTIC
480 YP=YMN+I*YTIC
490 LINE (XMN,YP)-(XMN+DXTIC,YP)
500 NEXT I
510 FOR I=1 TO NPTS-1
520 J=I+1
530 IF Y(I)>YMX OR Y(J)>YMX OR Y(J)<YMN OR X(I)<XMN
THEN 550
540 CIRCLE (X(I),Y(I)),0
550 NEXT I
560 FOR I=1 TO NXTIC
570 XP=XMN+XTIC*I
580 XC=PMAP(XP,0)
590 COL=INT(80*XC/640)-1
600 LOCATE 24,COL
```

```
610 PRINT USING "###.#"; XP;
620 NEXT I
630 FOR I=1 TO NYTIC
640 YP=YMN+I*YTIC
650 YR=PMAP(YP,1)
660 ROW=CINT(24*YR/199)+1
670 LOCATE ROW,1
680 PRINT USING "###.#"; YP
690 NEXT I
700 GOTO 700
```

In this program, I set the parameters σ, r, and b as follows:

$$\sigma = 10$$

$$r = 28$$

$$b = \tfrac{8}{3}$$

By selecting appropriate values for step size and the number of calculations per cycle (as discussed in Chapter 1 for the program SDEQ1) you can get a high degree of accuracy at the expense of CPU time. The usual picture of this three-dimensional strange attractor is shown in the XY, XZ, and YZ plots shown in Figs. 3-38, 3-39, and 3-40, respectively. The unusual oscillation for just the X-dimension is shown in Fig. 3-41. You might want to

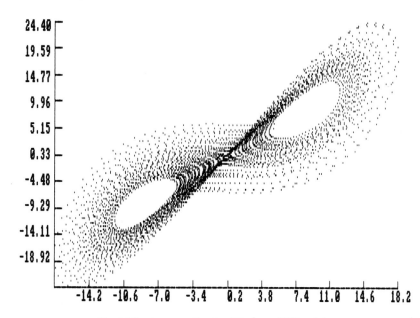

Fig. 3-38.   Lorenz attractor XY plane. 5000 points.

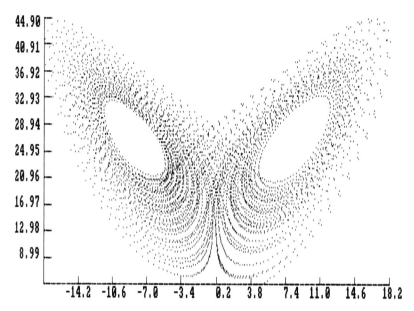

Fig. 3-39.   Lorenz attractor XZ plane. 5000 points.

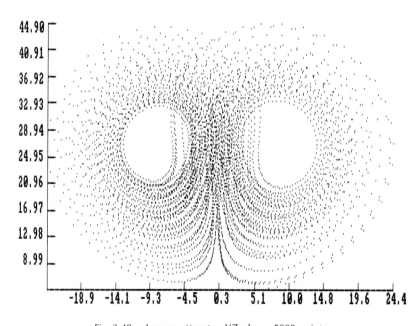

Fig. 3-40.   Lorenz attractor YZ plane. 5000 points.

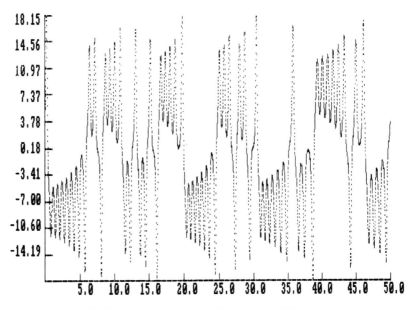

Fig. 3-41.   Lorenz attractor X-time plot. 5000 points.

experiment with changing the parameters σ, r, and b to see what other plots for the Lorenz attractor you produce.

Coupled-Logistic Map

Kaneko (1986) discusses the coupled-logistic map with a non-constant Jacobian. The map is given by the relation

$$x_{n+1} = 1 - Ax_n^2 + D(y_n - x_n)$$

$$y_{n+1} = 1 - Ay_n^2 + D(x_n - y_n)$$

and the Jacobian is

$$J = \begin{vmatrix} -2Ax_n + D & D \\ -2Ay_n + D & D \end{vmatrix}$$

The similarity of this and the logistic map of Chapter 2 can be seen with the following transform:

$$A = \lambda\left(\frac{\lambda}{4} - \frac{1}{2}\right)$$

$$x = \frac{(x' - 1/2)}{\left(\dfrac{\lambda}{4} - 1/2\right)}$$

$$y = \frac{(y' - 1/2)}{\left(\dfrac{\lambda}{4} - 1/2\right)}$$

We then get the following:

$$x'_{n+1} = \lambda x'_n(1 - x'_n) + D(y'_n - x'_n)$$

$$y'_{n+1} = \lambda y'_n(1 - y'_n) + D(x'_n - y'_n)$$

This map represents two logistic models coupled by a linear term.

The program KANEKO1 is a straightforward modification of ITEMAP2, which was introduced in Chapter 2. The maps show very interesting behavior for small changes in the parameter A with D held constant. In the following description, I let $D = 0.1$ and varied A.

KANEKO1

```
10 CLS
20 DIM X(1001),Y(1001)
30 PRINT "********  DATA FILE GENERATION PROGRAM *********"
40 PRINT "USED TO GENERATE ITERATED MAPS FILES"
50 INPUT "INPUT PARAMETER D";D
60 INPUT "INPUT PARAMETER A";A
70 INPUT "INPUT FILE NAME ";FILE$
80 OPEN "O",#1,FILE$
90 X(1)=.1
100 Y(1)=.15
110 FOR T=1 TO 1000
120 '     MAPPING EQUATIONS HERE
130 X(T+1)=1-A*X(T)*X(T)+D*(Y(T)-X(T))
140 Y(T+1)=1-A*Y(T)*Y(T)+D*(X(T)-Y(T))
150 X(I)=X(T+1)
160 Y(I)=Y(T+1)
280 PRINT #1,X(I),Y(I)
290 I=I+1
300  ' PRINT T,X(I),Y(I)
310 NEXT T
320 CLOSE #1
400 END
```

Figure 3-42 is a plot of the coupled logistic map with $A = 1.2$. Notice that this is a cycle of period two, designated 2T, which is analogous to a two-torus with two incommensurate frequencies. The analogy will become

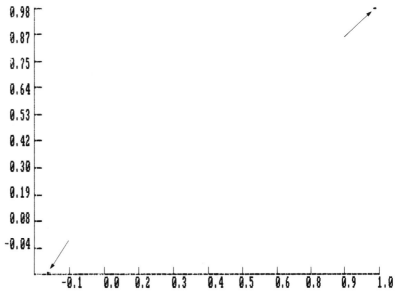

Fig. 3-42.   2T logistic map. D = 0.1, A = 1.2.

clearer later. As the parameter A is increased to $A = 1.3$, a period-four cycle appears as shown in Fig. 3-43. At $A = 1.35$, in Fig. 3-44 and Fig. 3-45, a period-eight cycle appears. In Fig. 3-46 ($A = 1.355$), and there appears to be some frequency locking, or *quasichaos*.

By frequency-locking, I mean the type of chaos bands that occur in phase-locked KAM tori. When $A = 1.373$, the frequency-locking has resulted in four islands of chaos, here referred to as 4C; this can be seen in Fig. 3-47. The eight islands in 8C have condensed into 4C. As A increases to $A = 1.40$, the 4C condenses into 2C as shown in Fig. 3-48. When $A = 1.55$, as can be seen in Fig. 3-49, hyperchaos is generated. To summerize, the route to chaos for the coupled logistic map is given schematically as follows:

$$2T \xrightarrow{\text{Bifurcate}} 4T \xrightarrow{\text{Bifurcate}} 8T \xrightarrow{\text{Phase locking}} 8C$$

$$\xrightarrow{\text{Phase locking}} 4C \xrightarrow{\text{Phase locking}} 2C \xrightarrow{\text{Fusion of chaos}} \text{Hyperchaos}$$

Two-Point Delayed Logistic Map

The two-point delayed logistic map has been studied by Kaneko (1986). This map is given by the following relation:

$$x_{n+1} = Ax_n + (1 - A)(1 - Dy_n^2)$$

$$y_{n+1} = x_n$$

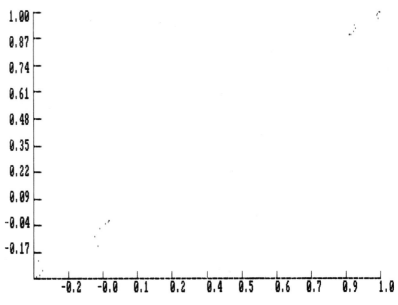

Fig. 3-43.   4T logistic map. D = 0.1, A = 1.3.

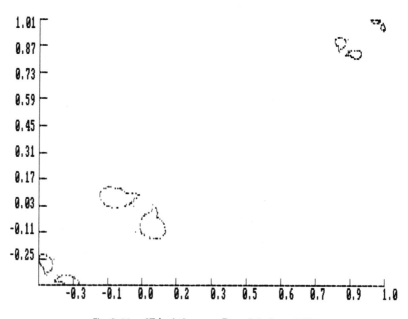

Fig. 3-44.   8T logistic map. D = 0.1, A = 1.35.

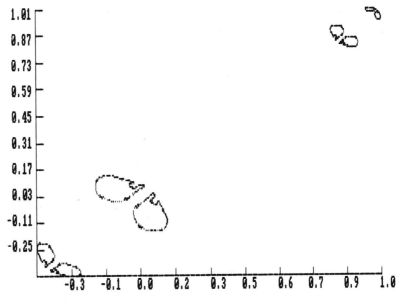

Fig. 3-45.   8T logistic map. D = 0.1, A = 1.352.

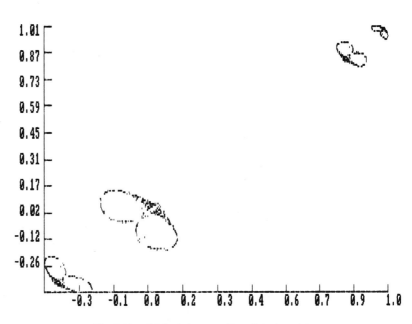

Fig. 3-46.   8C logistic map. D = 0.1, A = 1.355.

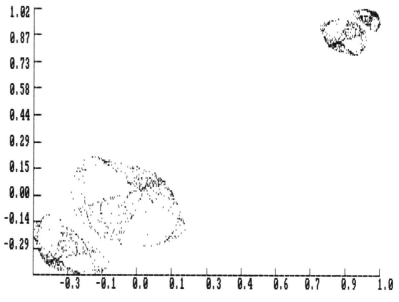

Fig. 3-47. 4C logistic map. D = 0.1, A = 1.373.

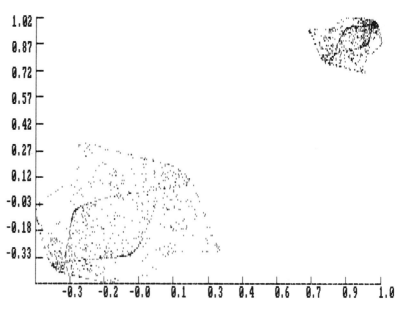

Fig. 3-48. 2C logistic map. D = 0.1, A = 1.4.

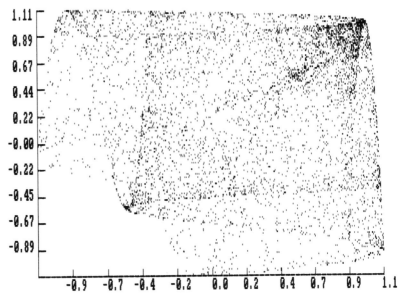

Fig. 3-49.   Hyperchaos logistic map. D = 0.1, A = 1.55.

The program KANEKO2 is a modification of KANEKO1. In this system, we will look at various values of *D* while holding *A* fixed.

KANEKO2

```
10 CLS
20 DIM X(1001),Y(1001)
30 PRINT "********  DATA FILE GENERATION PROGRAM *********"
40 PRINT "USED TO GENERATE ITERATED MAPS FILES"
50 INPUT "INPUT PARAMETER D";D
60 INPUT "INPUT PARAMETER A";A
70 INPUT "INPUT FILE NAME ";FILE$
80 OPEN "O",#1,FILE$
90 X(1)=.1
100 Y(1)=.15
110 FOR T=1 TO 1000
120 '     MAPPING EQUATIONS HERE
130 X(T+1)=A*X(T)+(1-A)*(1-D*Y(T)*Y(T))
140 Y(T+1)=X(T)
150 X(I)=X(T+1)
160 Y(I)=Y(T+1)
280 PRINT #1,X(I),Y(I)
290 I=I+1
300  ' PRINT T,X(I),Y(I)
310 NEXT T
320 CLOSE #1
400 END
```

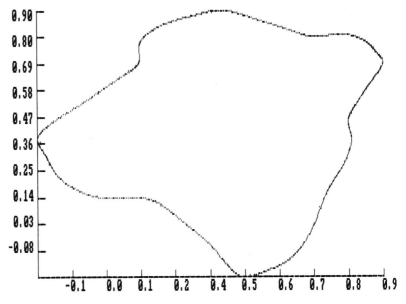

Fig. 3-50.   Delayed logistic map. D = 1.75, A = 0.3.

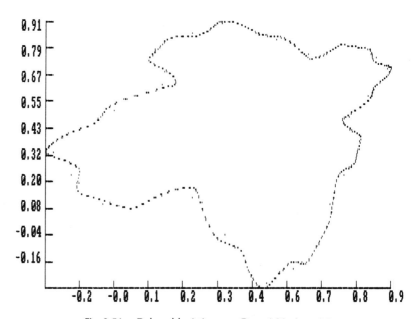

Fig. 3-51.   Delayed logistic map. D = 1.86, A = 0.3.

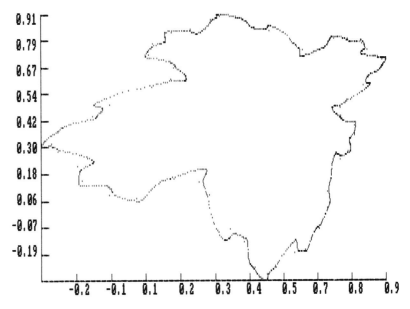

Fig. 3-52.   Delayed logistic map. D = 1.90, A = 0.3.

Figure 3-50 is a limit cycle or oscillation of a torus. This can be thought of as one band on a KAM torus. The parameter A is held fixed at $A = 0.3$. In Fig. 3-50 the parameter D is 1.75; in Fig. 3-51, $D = 1.86$. The oscillations begin to break up, perhaps as a result of phase-locking. When $D = 1.90$, there seems to be more phase-locking. This can be seen in Fig. 3-52. Figures 3-53 and 3-54 have a parameter of $D = 1.94$ and $D = 1.95$, respectively. These two orbits are bands of chaos in the KAM torus. When the parameter value reaches $D = 2.04$, the torus is almost broken up, as can be seen in Fig. 3-55. By the time $D = 2.16$, the torus has collapsed to hyperchaos, as can be seen in Fig. 3-56.

This and the previous example should convince you of the sensitive dependence of the critical parameter on the strange attractor.

Delayed Piecewise Linear Map

Kaneko (1986) simplified the two-point delayed logistic map and developed the delayed piecewise linear map of the following form:

$$x_{n+1} = Ax_n + (1 - A)(1 - D|y_n|)$$

$$y_{n+1} = x_n$$

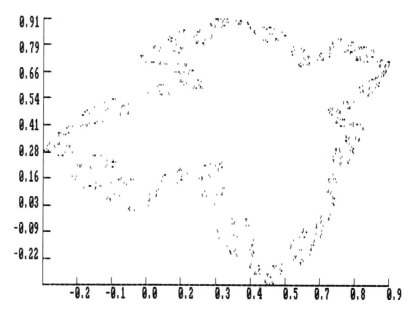

Fig. 3-53.   Delayed logistic map. D = 1.94, A = 0.3.

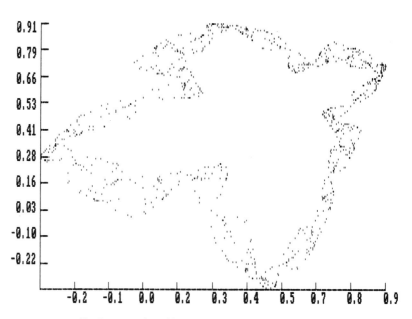

Fig. 3-54.   Delayed logistic map. D = 1.95, A = 0.3.

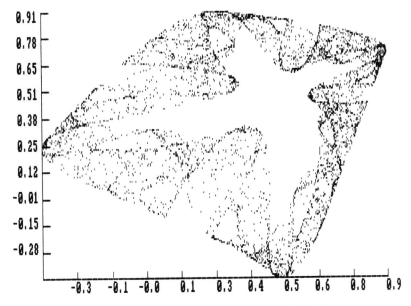

Fig. 3-55. Delayed logistic map. D = 2.04, A = 0.3.

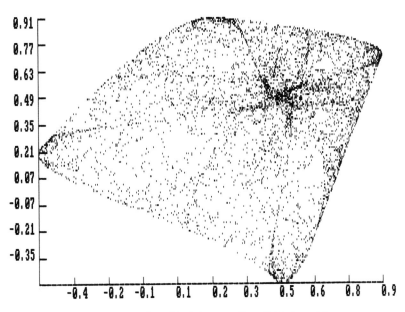

Fig. 3-56. Delayed logistic map. D = 2.16, A = 0.3.

The fixed point

$$(x,y) = \left(\frac{1}{(1 + D)}, \frac{1}{(1 + D)} \right)$$

is an unstable focus for $D > 1/(1-A)$, and chaos will appear immediately. For $y_n > 0$, the map reduces to a linear transform by rotation and stretching:

$$x'_{n+1} = Ax'_n - (1 - A)Dx'_n$$

$$y'_{n+1} = y'_n$$

where

$$x' = x - \frac{1}{(1 + D)}$$

$$y' = y - \frac{1}{(1 + D)}$$

When $y < 0$ the map is folded by the term $D|y|$.

This piecewise linear map has not been studied in much detail and is left for the experimenter. Here I will sketch out some of the ideas you might consider. You might want to modify the program KANEKO2 for this map. Set $A = 0.1$ and hold it fixed while changing D:

$$1.083 \leq D \leq 1.113$$

Then you can study the attractor of this map.

The Jacobian matrix for the map

$$x_{n+1} = Ax_n + (1 - A)(1 - D|y_n|)$$

$$y_{n+1} = x_n$$

has complex eigenvalues for $y > 0$

$$A \pm \sqrt{4c - A^2 i}$$

and for $y < 0$

$$A \pm \sqrt{4c + A^2 i}$$

where $C = D(1 - A)$. The Liapounov exponents are given by

$$\frac{1}{2} \log C$$

CATALOG OF STRANGE ATTRACTORS

All of the following strange attractors are from *Oscillation in Chemical Reactions* by Gurel and Gurel (1983). As you saw in the example of the Brusselator in this chapter, many chemical systems can exhibit strange attractors.

1. $$\frac{dx}{dt} = k_1A - k_2Bx + k_3x^2y - k_4x$$

    $$\frac{dy}{dt} = k_2Bx - k_3x^2y$$

2. $$\frac{dx}{dt} = -k_2x - k_3y\frac{x}{k + x} + k_1 + k_6z$$

    $$\frac{dy}{dt} = -k_2y - k_3x\frac{y}{k + x} + k_1 + b$$

    $$\frac{dz}{dt} = k_4y - k_5z$$

3. $$\frac{dx}{dt} = k_1 + k_2'x - (k_3y + k_4z)x/(x + k)$$

    $$\frac{dy}{dt} = k_5x - k_6y$$

    $$\frac{dz}{dt} = k_7x - k_8'z/(z + k')$$

4. $$\frac{dx}{dt} = -y - z - w$$

$$\frac{dy}{dt} = x$$

$$\frac{dz}{dt} = a(y - y^2) - bz$$

$$\frac{dw}{dt} = c\left(\frac{z}{2} - z^2\right) - dw$$

5. $$\frac{dx}{dt} = x(a_1 - k_1x - z - y) + k_2y^2 + a_3$$

$$\frac{dy}{dt} = y(x - k_2y - a_5) + a_2$$

$$\frac{dz}{dt} = z(a_4 - x - k_5z) + a_3$$

6. $$\frac{dx}{dt} = V_1 - b_1x - V$$

$$\frac{dy}{dt} = V_2 - b_2y + V - m(y - z)$$

$$\frac{dz}{dt} = ma(y - z)$$

$$V = \frac{x - ky}{1 + (x + y)(1 + cx^8)}$$

7. $$\frac{dx}{dt} = V_0 - b_1x - V$$

$$\frac{dy}{dt} = 1 - y - V$$

$$V = \frac{a \times y}{(1 + dy)(c + x(1 + x^8))}$$

4

Cellular Automata as Dynamical Systems

Cellular automata are discrete space-time models that can be used to model any system in the universe. They are a universe unto themselves. Cellular automata have been used to model biological systems from the level of cell activity to the levels of clusters of cells and populations of organisms. In chemistry, cellular automata have been used to model kinetics of molecular systems and crystal growth. In physics, they have been used to study dynamical systems as diverse as the interaction of particles and the clustering of galaxies. In computer science, cellular automata have been used to model parallel processing and von Neumann (self-reproducing) machines.

Cellular automata were invented in the late 1940s by J. von Neumann. Burke (1966) gives an excellent overview of the work by von Neumann. In cellular automata, space is divided into discrete small units called *cells* or *sites*. The sites take on a value, typically binary, of 0 or 1. At time t, all the cells will have a specific binary value. Rules local to a specific cell determine what the binary value of that cell will be at time $t + 1$.

Like space, time takes on discrete values. For the last several hundred years scientists have modeled the world with differential equations. Margolus (1984), Vichniac (1984), and Toffoli (1984) have shown that cellular automata are a good alternative to differential equations.

This chapter includes a discussion of cellular automata as dynamical systems. Many of the terms introduced in Chapters 2 and 3 are used to discuss cellular automata dynamics. Examples of local cellular automata rules are given with actual computer programs to simulate them. The final section of this chapter discusses cellular automata machines, and sets the stage for Chapter 5 on artificial neural networks, which are specialized cellular automata.

ATTRACTORS AND LIMIT CYCLES

The dynamics of cellular automata can be introduced in terms of a program called LIFE. This program is the framework used here to discuss attractors, limit cycles, and chaos. The game of LIFE was first introduced by John Conway, who has written about it at length in volume two of *Winning Ways for Your Mathematical Plays*, which he coauthored with Berlekamp, et al. (1982). Poundstone (1985) has also discussed LIFE at length in *The Recursive Universe* . . . . While describing the program line by line, I diverge to discuss various concepts and then return to the program description.

LIFE

```
10 DEF SEG =&HB800
20 DEFINT A-Y
30 SCREEN 0,0,0
40 RANDOMIZE TIMER
50 CLS
60 DIM A(25,80),D(25,80),E(25,80),HAMMING(30)
70 REM 219 IS ASCII FOR WHITE PIXEL
80 REM 255 IS ASCII FOR BLACK PIXEL
90 FOR I%=0 TO 24
100 FOR J%=0 TO 79
110 Z=RND(1)
120 IF Z<.1   THEN POKE I%*160+J%*2,219 ELSE POKE I%*160+J%
    *2,255
130 NEXT J%
140 NEXT I%
150 WHILE CYCLE < 30
160 CYCLE=CYCLE+1
170 FOR I%=0 TO 24
180 FOR J%=0 TO 79
190 B=PEEK(I%*160+J%*2)
200 IF B=219 THEN A(I%,J%)=219 ELSE A(I%,J%)=0
210 D(I%,J%)=A(I%,J%)
220 NEXT J%
230 NEXT I%
240 'INSERT CODE HERE FOR CELLULAR AUTOMATA RULES
250 FOR I%=1 TO 23
260 FOR J%=1 TO 78
270     C=A(I%-1,J%-1)+A(I%,J%-1)+A(I%+1,J%-1)
280     C=C+A(I%-1,J%)+A(I%+1,J%)
290     C=C+A(I%-1,J%+1)+A(I%,J%+1)+A(I%+1,J%+1)
```

```
300  IF C<=219*1 THEN    POKE I%*160+J%*2,255 : E(I%,J%)=0
310  IF C=219*3 THEN     POKE I%*160+J%*2,219 : E(I%,J%)=219
320  IF C>=219*4 THEN    POKE I%*160+J%*2,255 : E(I%,J%)=0
330 IF E(I%,J%)<>D(I%,J%) THEN 340 ELSE 350
340 HAMMING(CYCLE)=HAMMING(CYCLE)+1
350 NEXT J%
360 NEXT I%
370 LPRINT CYCLE;HAMMING(CYCLE)
380 WEND
390 END
```

Line 10 sets the current segment of memory to the screen buffer address for an AT&T PC6300 computer (an IBM XT clone). Line 20 defines integers, line 30 sets up the screen, line 40 randomizes the time, and line 50 clears the screen. After a DIM in line 60, the program begins with two nested loops starting in line 90. The screen is set up to be a 25 × 80 grid. ASCII values are POKEd into random places in this matrix. The density of ones can be adjusted by the parameter Z in line 120. If Z is less then 0.1 then the density is 0.1.

Characteristics of Cellular Automata

Before I discuss the WHILE loop in line 150, I'd like to diverge to discuss the fundamental characteristics of cellular automata. The first property is the geometry of the cell. In this program, the cell is rectangular. Because the grid is 25 × 80, there are 2000 cells total. A two-dimensional hexagonal array is possible, as is a three-dimensional array of cubic cells, or even a one-dimensional array.

Within a given array, it is necessary to specify the neighborhood that each cell examines in calculating its next state as it evolves in time from t to $t + 1$. When the state of neighbors at time t determines the state of a cell at time $t + 1$, it is said to be *local rules*. The two most common neighborhoods are the von Neumann and Moore, shown in Fig. 4-1.

In the von Neumann neighborhood, the cell (i,j) determines its state at time $t + 1$ based on the state of the four nearest neighbors. The Moore neighborhood uses the four diagonal neighbors also. The number of states per cell can be high. Von Neumann found a self-replicating pattern with cells of 29 states. Codd (1968) and Langton (1984) have discussed self-reproducing automata with only seven states per cell.

If k is the number of states per cell and n is the number of cells in the neighborhood then there are k^{k^n} possible rules. For example, consider a binary automata in a Moore neighborhood where $n = 8$. There are 10^{77}

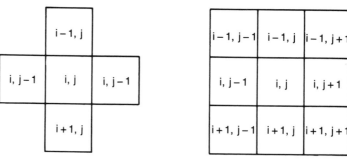

von Neumann Neighborhood   Moore Neighborhood

Fig. 4-1. The neighborhoods for cellular automata.

possible rules. The game of LIFE uses the Moore neighborhood. With 10^{77} possible rules, it can model the universe.

Continuing With the Program Description

In line 150, a WHILE loop starts for 30 cycles. These cycles are the time steps. All the cells change in parallel, with respect to a single time step, before the next time step. These time steps, or iterations, are called *cycles* in the program. Line 160 increments the cycle by one. The FOR-NEXT loops begin in lines 170 and 180 to PEEK into the screen address and write a value to the two-dimensional arrays *A* and *D*. The loops end in line 220 and 230.

The updating of each cell begins in lines 250 and 260. The actual rules are in lines 270 to 320. A variable, *C*, is assigned to the sum of the pixel values PEEKed and stored in the array *A*, as shown in Fig. 4-2. This results in summing the site value for the elements of the Moore neighborhood.

For the LIFE automata game, this sum is then checked with three threshold values. If the sum is less than or equal to one, then the cell *(i,j)*

```
270     C=A(I%-1,J%-1)+A(I%,J%-1)+A(I%+1,J%-1)
280     C=C+A(I%-1,J%)+A(I%+1,J%)
290     C=C+A(I%-1,J%+1)+A(I%,J%+1)+A(I%+1,J%+1)
300  IF C<=219*1 THEN    POKE I%*160+J%*2,255 : E(I%,J%)=0
310  IF C=219*3  THEN    POKE I%*160+J%*2,219 : E(I%,J%)=219
320  IF C>=219*4 THEN    POKE I%*160+J%*2,255 : E(I%,J%)=0
```

Fig. 4-2. The LIFE algorithm.

will take a zero state at time $t + 1$. That is, the cell will die from exposure. If the sum is equal to three, the (i,j) cell has three neighbors, each with a state value of one. The cell (i,j) will then take on the value of one. In the LIFE game, this is said to be a birth. Three cells generate the birth of a fourth cell. If the cell (i,j) already has a state value of one, then this value is maintained. If the sum is greater than or equal to four, then the cell (i,j) will take on a state value of zero at time $t + 1$. The cell is said to die from overcrowding.

In the Moore neighborhood, the maximum value the sum can obtain is eight, because the cell (i,j) has eight neighbors and we are dealing only with binary states. In the von Neumann neighborhood, the maximum sum value is four, because there are only four neighbors. After these threshold decisions take place, an element in a small array called *HAMMING(cycle)* is incremented if a cell change has taken place. This element called HAMMING requires some explanation.

The Hamming Distance

Suppose you are given two binary vectors of equal length:

$$A = (1\ 0\ 0\ 1\ 1\ 0)$$

$$B = (1\ 1\ 0\ 1\ 0\ 0)$$

These two vectors can be compared on a bit by bit basis. The number of differing bits is called the *Hamming distance*. For this example, the two vectors differ by two bits. So the Hamming distance is said to be two.

As you would expect when the cellular automata begins its processing, or updating, at $t = 1$ there are a greater number of changes than at $t = 30$. Plots of the iteration time, or number of cycle updates, versus Hamming distance can show the activity of the cellular automata network. This activity relation can be thought of as an *entropy*. At high entropy, there is a higher Hamming distance. At low entropy, there is a low Hamming distance.

Figure 4-3 shows a run of the program. This figure includes two limit cycles, two attractor points, and three regions of high entropy. By the terminology of LIFE, the two attractor points are known as *beehives* and the two limit cycles, both of period two, are known as *blinkers*. In order to keep the program short and simple to discuss cellular automata, I did not consider edge effects. This causes stagnation at the edges. Some programs do a wrap around to make the screen a torus.

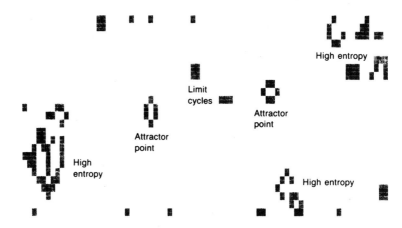

Fig. 4-3.   Screen dump from LIFE.

Figures 4-4 through 4-6 show runs for three different density values. After twenty iterations, Fig. 4-4 shows three attractor points and two limit cycles. The corresponding Hamming plot for this system is shown in Fig. 4-7. Notice that after about nine iterations, the Hamming distance has settled down to a constant. Figure 4-5 shows a system after fifty iterations,

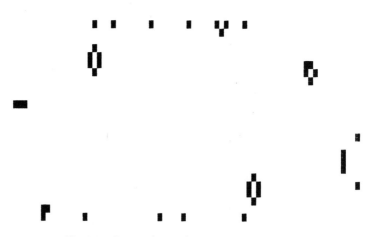

Fig. 4-4.   Screen dump after 20 iterations. Z = 0.1.

Fig. 4-5. Screen dump after 50 iterations. Z = 0.15.

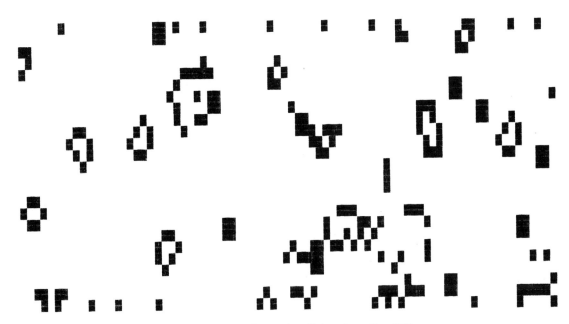

Fig. 4-6. Screen dump after 50 iterations. Z = 0.20.

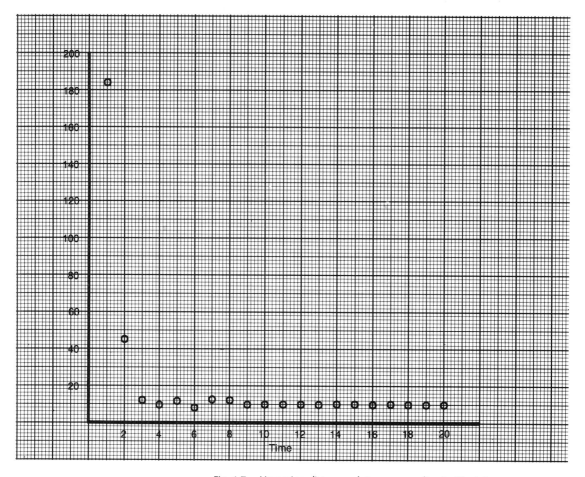

Fig. 4-7. Hamming distance plot corresponding to Fig 4-4.

starting with an initial density of $Z = 0.15$. There is a high degree of entropy, as can be seen in the Hamming plot of Fig. 4-8. Also from Figure 4-5 you can see some attractor points and limit cycles.

Fig. 4-6 is a system configuration after fifty iterations and Fig. 4-9 is the corresponding Hamming plot for an initial density of $z = 0.20$. The high entropy region can be considered to be chaos at this time in evolution. But as the system evolves, the chaos gives limit cycles and attractor points.

The LIFE Algorithm

Before closing this section and moving on to a discussion of entropy and Liapounov exponents, I would like to give a mathematical summary

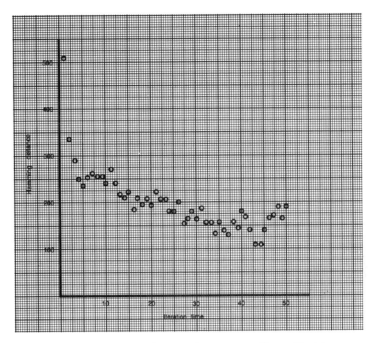

Fig. 4-9.   Hamming distance plot corresponding to Fig 4-6.

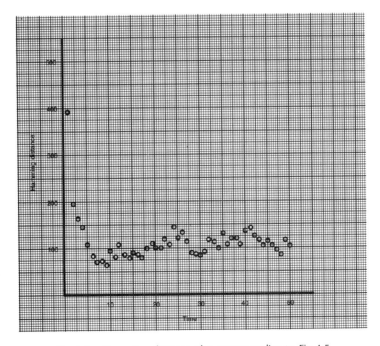

Fig. 4-8.   Hamming distance plot corresponding to Fig 4-5.

of the LIFE alogrithm. The LIFE algorithm can be generalized as follows:

$$\delta(a_0, a_1, \ldots, a_8) = \begin{cases} 1 & \text{if} \begin{cases} a_0 = 1 \text{ and } 2 \leq \sum_{i=1}^{8} \leq 3 \\ \text{or} \\ a_0 = 0 \text{ and } \sum_{i=1}^{8} a_i = 3 \end{cases} \\ 0 & \text{otherwise} \end{cases}$$

From this alogrithm you can see how to construct other rules. Furthermore, this type of algorithm and notation is used to describe the behavior of artificial neural networks in Chapter 5.

It might be interesting to change the threshold rule in this program to a random integer between one and eight to produce a stochastic rule. Another interesting experiment would be to allow different sites to follow different update rules, thus creating heterogenous cellular automata.

ENTROPY AND LIAPOUNOV FUNCTIONS

Cellular automata, like LIFE show self-organization out of chaos. Fixed attractor points and limit cycles are the end result of many iterations. These self-organizing systems appear to violate the Second Law of thermodynamics. They actually circumvent this Law the same way other dynamical systems such as strange attractors do—by shrinking in time the volume element of the phase space. The dynamics are irreversible in time. Any given configuration can have a large number of paths that lead to this one configuration. Symmetric structures often arise as a result of irreversibility. And symmetric structures can have symmetric and asymmetric parents.

Recall from earlier chapters that a negative Liapounov exponent is a measure of the rate of convergence of different initial conditions toward a common attractor or fixed point. The amount of merging in a binary system can be measured by the following:

$$S^{(N)}(t) = -\sum_{x=0}^{2^n-1} P^{(N)}(X,t) \log P^{(N)}(X,x)$$

In this expression, $P^{(N)}(X,t)$ is the probability that a configuration $x = (x_1, x_2, \ldots, x_N)$ is reached at time t, where N is the number of sites. This relation is also the definition of entropy. The synchronous parallel updating makes this a decreasing function in time:

$$S^{(N)}(t + 1) \leq S^{(N)}(t)$$

In order to discuss the Liapounov exponent in more detail, I must first introduce the classes of cellular automata. Wolfram (1983, 1985) has done extensive analysis on one-dimensional cellular automata and has discovered four classes of them. All four classifications are based on the limiting configuration after many iterations. In the first class, all sites ultimately attain the same value. In the second class, simple stable or periodic separated structures are formed. The LIFE rules are an example of this class. In the third class chaotic patterns are formed, such as strange attractors. In the fourth class, complex localized structures are formed.

The sensitive dependence on initial conditions can easily be observed. Start with a randomly chosen initial configuration and let the system evolve for a large number of iterations, say M iterations. Observe the resulting configuration. Now go back to the same random initial configuration and change one cell and let the system evolve for M iterations again. Class I automata show no effect. Class II automata might show a very small effect confined to a small region near the site of the change. Class III automata show a large effect, just as would be expected for a strange attractor. Class IV automata are so rare and unpredictable that the best way to predict the outcome is to allow the cellular automata to compute the final state.

All of Wolfram's analyses were based on one-dimensional cellular automata. His rules for one-dimensional celluiar automata are as follows: Each binary cell assumes one of two values at each iteration. The output state is a binary digit, which is determined by previous states of three binary digits. The rule may be thought of as a three-input binary logic gate. Because there are eight possible input combinations, there are $2^8 = 256$ possible iteration rules.

Each rule may be expressed as an eight-digit binary number. From these 256 rules, Wolfram deduced 32 legal rules. A legal rule is one which is reflection-symmetric and under which the state containing all zeros is stable. This eliminates many systems. Wolfram has studied these 32 legal rules intensively for $k = 2$ and $r = 1,2,3$ where k is the number of states per site—in other words, binary cells with the range r as the number of nearest neighbors.

Wolfram's results are summarized in Table 4-1. This table shows the fraction of legal cellular automata in each of the four basic classes. You can clearly see that Class IV is very rare.

Table 4-1

Class of Automata	k = 2 r = 1	k = 2 r = 2	k = 2 r = 3	k = 3 r = 1
I	0.50	0.25	0.09	0.12
II	0.25	0.16	0.11	0.19
III	0.25	0.53	0.73	0.60
IV	0.00	0.06	0.60	0.07

From Chapter 3 you know that strange attractors have a positive Liapounov exponent, and limit cycles have a zero exponent. Packard (1985) has calculated the Liapounov exponents for most of the legal rules of Wolfram's one-dimensional cellular automata. Packard has found that all Class I and II cellular automata have a zero Liapounov exponent. Class III automata have a positive Liapounov exponent, as you would expect for strange attractors. Some of the Class III cellular automata Liapounov exponents are given in Table 4-2. All class III cellular automata have a positive Liapounov exponent for all initial conditions.

Table 4-2

Class III Rule	Liapounov Exponent
90	1.0
18	0.99
193	0.5
86	0.98
22	0.82

REVERSIBLE CELLULAR AUTOMATA

A reversible cellular automata can be followed in reverse, after M time steps, to its initial configuration. Margolus (1984) has studied reversible cellular automata and Fredkin and Toffoli (1982) have studied reversible logic. Any cellular automata can be described by the relation

$$S_{i,t+1} = f(S_{i,t})$$

where $S_{i,t+1}$ is the state of cell i at time $t + 1$ and $f(S_{i,t})$ is a function of the cells in the neighborhood of i at time t. Given the following relation,

$$S_{t+1} = f(S_t) + S_{t-1}$$

the function will be reversible when the following is true:

$$S_{t-1} = f(S_t) - S_{t+1}$$

Any function f that follows this relation will be reversible.

Now let's look at a program that follows Fredkin logic. The program FREDKIN is a modification of the program LIFE. Only the cellular automata rules have been changed. In the Fredkin cellular automata, a cell will be on in the next iteration if and only if one or three of its four von Neumann neighbors are presently on. If zero or two of its neighbors are on, the cell

FREDKIN

```
10 DEF SEG =&HB800
15 DEFINT A-Y
20 SCREEN 0,0,0
30 CLS
40 DIM A(25,80),B(25,80)
50 REM 219 IS ASCII FOR WHITE PIXEL
60 REM 255 IS ASCII FOR BLACK PIXEL
70 FOR I%=0 TO 24
80 FOR J%=0 TO 79
100 IF I%=12 AND J%=40 THEN POKE I%*160+J%*2,219 ELSE POKE
    I%*160+J%*2,255
120   POKE   I%*160+J%*2+1,10
140 NEXT J%
150 NEXT I%
160 FOR I%=0 TO 24
170 FOR J%=0 TO 79
180 B=PEEK(I%*160+J%*2)
190 IF B=219 THEN A(I%,J%)=219 ELSE A(I%,J%)=0
210 NEXT J%
211 NEXT I%
215 'INSERT CODE HERE FOR CELLULAR AUTOMATA RULES
220 FOR I%=1 TO 23
230 FOR J%=1 TO 78
260    C=A(I%,J%-1)
270    C=C+A(I%-1,J%)+A(I%+1,J%)
280    C=C+A(I%,J%+1)
292  IF C=219 OR C=219*3 THEN POKE I%*160+J%*2,219 ELSE
    POKE I%*160+J%*2,255
300 POKE I%*160+J%*2+1,10
400 NEXT J%
401 NEXT I%
402 CYCLE = CYCLE + 1
450 GOTO 160
1000 END
```

will be off in the next generation. This logical rule gives rise to self-repro-ducing cellular automata.

When you run the program FREDKIN, you can see that the center cell has a state value of one. This cell will continue to reproduce, as shown in Fig. 4-10. Notice that at even times, the cells in state one are not touching at their corners, while at odd times they are. This even/odd time result is a consequence of the even/odd rules for the state of the cellular automata.

Fig. 4-10. Fredkin logic screen dumps.

Another related cellular automata program is CELL1. This program uses bit graphics rather than POKE and PEEK graphics, and is therefore much faster. The cellular automata rules for this program are random at each iteration. The program starts with the cell in the upper-left corner active.

CELL1

```
10  KEY OFF
20  CLS
30  CLEAR
40  DEFINT A-Z
50  SCREEN 1
60  P=8000
70  DIM A(P)
80  PSET(1,1)
90  GET (1,1)-(318,198),A
100 XI=SGN(RND-.5)
110 YI=SGN(RND-.5)
120 PUT(1+XI,1+YI),A
130 GOTO 90
```

The cellular automata rules, even though they are random, give rise to a reasonable degree of self-generation, as can be seen in Fig. 4-11. This figure, after scores of iterations, shows self-reproduction on the advancing front. The next section discusses the fractal nature of cellular automata.

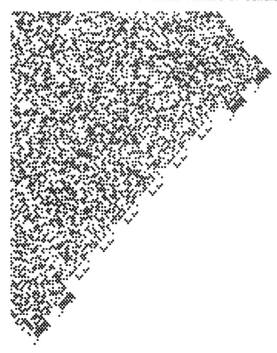

Fig. 4-11.   Screen dump from program CELL1.

LATTICE ANIMALS AND FRACTALS

In this section I discuss what are known as stochastic models of cluster growth. With the appropriate cellular automata rules, clusters and aggregates evolve after many iterations. These cluster models have applications in modeling the formation of microparticles such as metallic aggregates, soot, and smoke. They can also be used in modeling two-phase flow or perculation and electric discharge in solids.

Vannimenus et al. (1985) have shown that many clusters formed from very small particles do not behave like ordinary matter. In these microclusters, the density goes to zero as the size increases. The only way for this to happen is if cluster growth is fractal. The number N of the constituent particles and the size of the clusters R is given by the scaling law

$$N \approx AR^D$$

where D is the fractal dimension. If D is less than the space dimension, then the average density goes to zero for large N. This particle growth can be modeled with cellular automata and the fractal dimension can be deduced.

There are several types of aggregation models. One class, called *lattice animals*, consists of all types of connected graphs. *Growing animals*, or *Eden models*, are a second class. These are called growing animals because new particles are added at random on the boundary sites. If particles are allowed to diffuse randomly before sticking to the growing cluster or leaving to infinity, the model is called *diffusion-limited-aggregation* (DLA). Another type of cluster is *clustering of clusters*, like the formation of galactic superclusters. Vannimenus et al. (1985) have calculated the fractal dimension for these four models of cluster growth. Their results are summarized in Table 4-3. These results show that the effects of kinetic growth significantly modify the fractal dimension.

Table 4-3

d	Lattice Animals	Eden Model	Diffusion	Superclusters
2	1.56	2	1.7	1.4
3	2	3	2.4	
4	2.4		3.3	

Another interesting class of cluster growth is *directed clusters*, or directed aggregation. The program DIRECT was written from a simple modification of the program LIFE. This program has the simple cellular automata rule that if one of the upper or lower nearest neighbors to cell *i* has a state value of one at time *t*, then cell *i* will have a state value of one at time $t + 1$.

DIRECT

```
10 DEF SEG =&HB800
20 DEFINT A-Y
30 SCREEN 0,0,0
40 RANDOMIZE TIMER
50 CLS
60 DIM A(25,80),D(25,80),E(25,80),HAMMING(30)
70 REM 219 IS ASCII FOR WHITE PIXEL
80 REM 255 IS ASCII FOR BLACK PIXEL
90 FOR I%=0 TO 24
100 FOR J%=0 TO 79
110 Z=RND(1)
120 IF Z<.1   THEN POKE I%*160+J%*2,219 ELSE POKE I%*160+J%
    *2,255
130 NEXT J%
140 NEXT I%
150 WHILE CYCLE < 1000
160 CYCLE=CYCLE+1
170 FOR I%=0 TO 24
180 FOR J%=0 TO 79
190 B=PEEK(I%*160+J%*2)
200 IF B=219 THEN A(I%,J%)=219 ELSE A(I%,J%)=0
210 D(I%,J%)=A(I%,J%)
220 NEXT J%
230 NEXT I%
240 'INSERT CODE HERE FOR CELLULAR AUTOMATA RULES
250 FOR I%=1 TO 23
260 FOR J%=1 TO 78
270 IF A(I%-1,J%)=219 OR A(I%+1,J%)=219 THEN 280 ELSE 290
280 POKE I%*160+J%*2,219
285 GOTO 350
290 POKE I%*160+J%*2,255
350 NEXT J%
360 NEXT I%
380 WEND
390 END
```

Figure 4-12 shows a run of this program after twenty-eight iterations. The simulation starts with a random initial configuration. The self-organization behavior of this cellular automata is very evident. This could be called a directed lattice animal cellular automata.

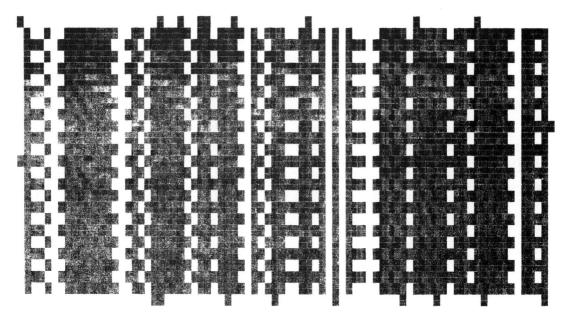

Fig. 4-12. Screen dump of a directed cluster from the program DIRECT.

Although the program LIFE does not consider boundary effects and it is slow, its major advantage is that it is easy to modify, as this example shows, to study other cellular automata rules.

CELLULAR AUTOMATA COMPUTERS

As stated earlier, cellular automata are essentially parallel-processing computing units. The initial condition is the data input, the cellular automata rules are the program, and the final configuration is the computed result. Each iteration is a clock cycle. The parallel processing can be synchronous or asynchronous, depending on the updating procedure during each iteration. However, this is a rather abstract computer. The more conventional computer made from AND, OR, and NOT gates can also be simulated with a cellular automata such as the program LIFE. Berlekamp et al. (1982), Poundstone (1985) and Dewdney (1985) have given excellent descriptions of how to construct a computer with the LIFE algorithm. Poundstone (1985) has gone into great detail on the construction, starting from simple logic gates and building up to a computer that can reproduce itself. These self-reprodcing computers are called *von Neumann machines*.

Several other cellular automata computers have been described in the literature. Margolus (1984) has described a billiard ball model based on reversible cellular automata, and Carter (1984) has described molecular-scale computers built on the principle of molecular engineering and cellular automata.

Toffoli (1984) was tired of watching slow cellular automata evolve on a computer screen. He decided to build a special-purpose cellular automata machine. The entire unit consisted of a black box with some logic boards interfaced to a VIC-20 computer and a color monitor. Later, Toffoli and Margolus (1987) described extensive research that they conducted with this and other cellular automata machines. Hillis (1985) has described a large cellular automata machine that he calls the Connection Machine.

In the next chapter, I discuss a special class of cellular automata known as neural networks. I also discuss some of the complex dynamics of neural networks.

5

The Dynamics of Neural Networks

Chapter 4 examined cellular automata. These cellular automata can be used to model parallel processing such as the parallel processing in a neural network. Each site in the cellular automata acts as one neuron in the neural network. The neurons, however, do not just interact with nearest neighbors, but can interact with all other neurons in the network.

In this chapter, I discuss artificial neural networks and show some of their computational properties. In particular, I show that there are stable states, or attractor points, in a neural network, and that these attractor points are fixed memories. I also show how simple limit cycles can arise from neural computation and discuss how overlapping memory states can generate spurious attractor states. These stable states and limit cycles arise as a result of the content-addressable nature of the neural networks. The discussion begins, therefore, with an extensive discussion of the mathematics of parallel processing and content addressing.

THE MATHEMATICS OF PARALLEL DISTRIBUTED PROCESSING

This section introduces the mathematical methods and fundamental theoretical concepts of parallel distributed processing using threshold logic devices. The basic concepts of model neural networks can be described by vector analysis and linear algebra.

A *matrix* is an array of elements of real numbers. For example, consider the following:

$$M = \begin{bmatrix} 3 & 2 & 9 \\ 7 & 6 & 0 \\ 1 & 4 & 2 \end{bmatrix}$$

M is a three-dimensional matrix, or a 3 × 3 matrix. Matrices need not be square, for example:

$$P = \begin{bmatrix} 2 & 0 \\ 7 & 1 \\ 5 & 4 \end{bmatrix} \qquad N = \begin{bmatrix} 1 \\ 0 \\ 1 \end{bmatrix}$$

P is a 3 × 2 matrix and N is a 3 × 1 matrix.

It is sometimes convenient to think of a vector as a one-dimensional matrix:

$$V = (3\ 1\ 0)$$

Multiplication of a matrix by a scalar is the same as multiplication of a vector by a scalar. Each element in the matrix is multiplied by the scalar:

$$3*P = \begin{bmatrix} 3{\cdot}2 & 3{\cdot}0 \\ 3{\cdot}7 & 3{\cdot}1 \\ 3{\cdot}5 & 3{\cdot}4 \end{bmatrix} = \begin{bmatrix} 6 & 0 \\ 21 & 3 \\ 15 & 12 \end{bmatrix}$$

Addition of matrices is similar to addition of vectors, for example, if

$$M1 = \begin{bmatrix} 1 & 0 & 5 \\ 0 & 7 & 2 \\ 4 & 6 & 6 \end{bmatrix} \qquad M2 = \begin{bmatrix} 6 & 9 & -3 \\ 5 & 2 & 0 \\ -8 & 4 & 4 \end{bmatrix}$$

then

$$M1 + M2 = \begin{bmatrix} 1+6 & 0+9 & 5-3 \\ 0+5 & 7+2 & 2-0 \\ 4-8 & 3+4 & 6+4 \end{bmatrix} = \begin{bmatrix} 7 & 9 & 2 \\ 5 & 9 & 2 \\ -4 & 7 & 10 \end{bmatrix}$$

This has application in memory storage. If each matrix represents one memory, then the sum of the two matrices results in a storage matrix for the two memory states.

A very important concept is the multiplication of a vector by a matrix. This can be used in pattern recognition and memory recall. For example, given a vector

$$v = \begin{bmatrix} 2 \\ 9 \\ 7 \end{bmatrix}$$

and a matrix

$$M = \begin{bmatrix} 1 & 0 & 5 \\ 0 & 7 & 2 \\ 4 & 3 & 6 \end{bmatrix}$$

the inner product is found as follows:

$$u = Mv = \begin{bmatrix} 1 & 0 & 5 \\ 0 & 7 & 2 \\ 4 & 3 & 6 \end{bmatrix} \begin{bmatrix} 2 \\ 9 \\ 7 \end{bmatrix}$$

$$u = \begin{bmatrix} 1 \cdot 2 + 0 \cdot 9 + 5 \cdot 7 \\ 0 \cdot 2 + 7 \cdot 9 + 2 \cdot 7 \\ 4 \cdot 2 + 3 \cdot 9 + 6 \cdot 7 \end{bmatrix} = \begin{bmatrix} 37 \\ 77 \\ 77 \end{bmatrix}$$

Notice that the inner product of a matrix with a vector is a vector. The matrix need not be square, as shown in the following example:

$$K = \begin{bmatrix} 3 & 7 & 1 \\ 0 & 1 & 2 \end{bmatrix} \qquad t = \begin{bmatrix} 1 \\ 0 \\ 3 \end{bmatrix}$$

$$q = Kt = \begin{bmatrix} 3 & 7 & 1 \\ 0 & 1 & 2 \end{bmatrix} \begin{bmatrix} 1 \\ 0 \\ 3 \end{bmatrix}$$

$$= \begin{bmatrix} 3 \cdot 1 + 7 \cdot 0 + 1 \cdot 3 \\ 0 \cdot 1 + 1 \cdot 0 + 2 \cdot 3 \end{bmatrix} = 6 \begin{bmatrix} 1 \\ 1 \end{bmatrix}$$

This is sufficiently important to be written in symbolic terms:

$$u = Mv$$

$$i^{th} \text{component of } u = (i^{th} \text{ row of } m)(v)$$

A convenient way of thinking of this operation is as a mapping, where v is mapped to u by the operation M (see Fig. 5-1).

This mapping of one state into another is analogous to a one-layer parallel distributed processing system. In Fig. 5-2, there are n input units and p output units. Each input processor is connected to each output

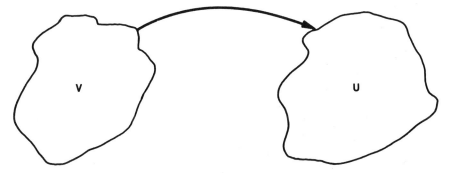

Fig. 5-1. Map of v to u by process M.

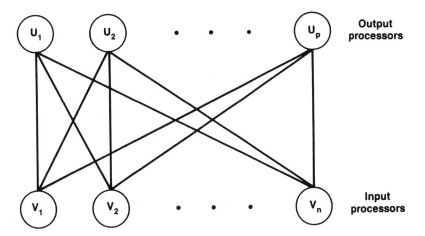

Fig. 5-2. Two layer process. All processors at v are connected to all processors at u.

processor by a connection strength, M_{pn}, where M_{pn} represents the pn^{th} element in the matrix M. Each output unit computes the inner product of its weight vector and the input vector. In other words, the output at the ith output processor is found by computing the inner product of the input vector with the weight vector for the ith processor. The components of the input vector are the values of the input units: The weight vector for the ith process is the ith row of the strength matrix M.

This technique can be extended to multilayered systems where the output of one layer becomes the input of the next layer. In Fig. 5-3, the processors at a are connected to each of the processors at b. These are in turn connected to each of the c processors. An input vector at a is mapped to b by the connection strength matrix Y. Vector b is then mapped to vector c by connection strength matrix Z. This can be represented

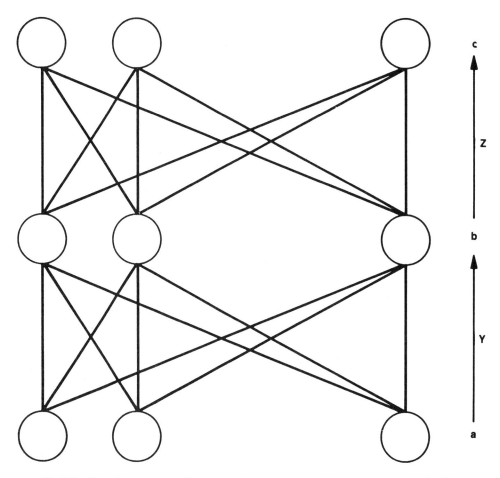

Fig. 5-3. Three layer process. Processors at a are connected to processors at b, by the matrix, Y. Processors at b are connected to those at c, by matrix Z.

symbolically as follows:

$$c = Z(Ya).$$

In other words, the matrix vector product of Ya results in the vector b, and the matrix vector product Zb results in c.

A very important mathematical technique in parallel distributed-processing theory is the transpose and outer product of a vector. The transpose of a $wn \times m$ matrix is a $wm \times n$ matrix. If the matrix is a one-dimensional matrix, i.e., a vector, then the transpose is found as demonstrated in the following example. Given a vector

$$v = (3\ 7\ 9\ 2)$$

then the transpose of v is given by the following:

$$v^t = \begin{bmatrix} 3 \\ 7 \\ 9 \\ 2 \end{bmatrix}$$

Notice the superscript t. This indicates the transpose operation.

The inner product of a vector with a transposed vector gives a scalar, as shown in the following example:

$$v = (9\ 7\ 3)$$

$$u = (5\ 2\ 0)$$

$$vu^t = [9\ 7\ 3] \begin{bmatrix} 5 \\ 2 \\ 0 \end{bmatrix} = 9 \cdot 5 + 7 \cdot 2 + 3 \cdot 0 = 59$$

The outer product of a vector with a transposed vector gives a matrix:

$$uv^t = [5\ 2\ 0] \begin{bmatrix} 9 \\ 7 \\ 3 \end{bmatrix} = \begin{bmatrix} 45 & 18 & 0 \\ 35 & 14 & 0 \\ 15 & 6 & 0 \end{bmatrix}$$

The outer product concept can be applied to learning in a neural network. This is called the Hebb learning rule (Rumelhart, et al. 1986). A particular matrix can be generated by associating an input vector with an output vector. This is known as *associative learning*. This technique is used in an example program later in this chapter.

For any given vector v, when the outer product of v with its transpose v^t is found, a memory matrix unique for that memory state is generated. If the inner product of this memory state and the memory matrix is found, then the result is the memory. The input vector need not be the pure memory state, but only a partial memory. When this partial memory state is operated on by the memory matrix, the inner product will give the complete and correct memory state.

What is not obvious from the above is that a given memory matrix can store more than one memory state. The actual number of memories depends on the size of the matrix.

THRESHOLD LOGIC AND NONLINEAR SYSTEMS

The human brain is a massive parallel personal computer based on organic threshold logic devices. These logic devices are known as *neurons*. A neuron is sketched in Fig. 5-4.

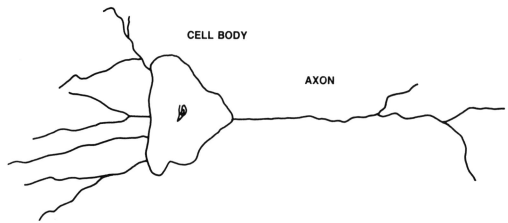

DENDRITES

CELL BODY

AXON

Fig. 5-4.   Simplified diagram of a neuron.

The cell body has one or more output lines called the *axon*. The input lines are called *dendrites*. There are about 1000 dendrite connections to an average neuron in the human brain. The neurons are interconnected by a *synapse*. A neuron can have both excitatory and inhibitory connections. An excitatory connection tells the neuron to fire, and an inhibitory connection tells a neuron to not fire. The input signals are summed in the neuron. At a certain threshold level, the neuron will fire; below this level, it will not fire. This is diagrammed in Fig. 5-5.

As shown in Fig. 5-5, if the sum of the input signal I_{in} is less than threshold I_t, then V_{out} goes low, represented by logic 0. If the sum of input signal I_{in} is above threshold I_t, then V_{out} goes high, logic 1. This idea can be represented algebraically as follows:

$$V_{out} = \begin{cases} 1 \\ 0 \end{cases} \text{ if } I_{in} \quad \begin{matrix} > I_t \\ < I_t \end{matrix}$$

Software implementation of this idea is presented later in this chapter.

This model is a little naive, because the synapse connections to the inputs of the threshold logic device have not been considered. In the hardware implementation of threshold logic circuits, the synapse is a

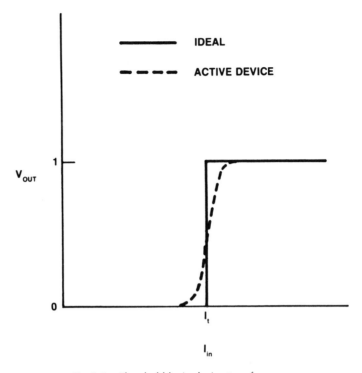

Fig. 5-5.   Threshold logic device transfer curve.

resistive interconnection between logic devices. A less naive model in algebraic terms is given below:

$$V_{out} = \begin{cases} 1 \\ 0 \end{cases} \quad \text{if} \sum_{j \neq 1} w_{ii}v_j \quad \begin{matrix} > I_t \\ < I_t \end{matrix}$$

In this relation, the V_j is the input signal to neuron j and W_{ij} is the conductance of the connection between the ith and j^{th} neurons. Each threshold logic unit randomly and asynchronously computes whether it is above or below the threshold and readjusts accordingly. Therefore, a network of these threshold logic units is a parallel computer.

This parallel computer can be used in optimization problems and for content-addressable memories. The content-addressable memory implementation of these parallel computation circuits is discussed at length in this chapter.

The information storage algorithm for content-addressable memories, discussed in this chapter, is called the Hebb learning rule. A memory state is given as a binary vector. In a binary vector, all the elements are either

1 or 0. The outer product of this memory vector with its transpose gives a storage matrix.

Let's take an example. The ASCII code for the letter A is given by the standard binary representation [0 1 0 0 0 0 0 1]. The outer product of this eight-dimensional binary vector with its transpose is given below:

$$
\begin{bmatrix} 0 \\ 1 \\ 0 \\ 0 \\ 0 \\ 0 \\ 0 \\ 1 \end{bmatrix} [0\ 1\ 0\ 0\ 0\ 0\ 0\ 1] =
\begin{bmatrix}
0 & 0 & 0 & 0 & 0 & 0 & 0 & 0 \\
0 & 1 & 0 & 0 & 0 & 0 & 0 & 1 \\
0 & 0 & 0 & 0 & 0 & 0 & 0 & 0 \\
0 & 0 & 0 & 0 & 0 & 0 & 0 & 0 \\
0 & 0 & 0 & 0 & 0 & 0 & 0 & 0 \\
0 & 0 & 0 & 0 & 0 & 0 & 0 & 0 \\
0 & 0 & 0 & 0 & 0 & 0 & 0 & 0 \\
0 & 1 & 0 & 0 & 0 & 0 & 0 & 1
\end{bmatrix}
$$

This storage algorithm can be represented in algebraic terms as follows:

$$W = m^t m$$

W is the information storage matrix, m is the memory vector and m^t is the transpose of vector m.

This model does not consider an important point. The elements W_{ii} should be set to zero. Hopfield (1982, 1984) and McEliece, et al. (1985) have shown that if W_{ii} is not zero, then the hardware implementation of the model can result in chaotic oscillations. The correct algebraic relation is

$$W = m^t m - I_n.$$

where I_n is the $n \times n$ identity matrix. So the storage algorithm consists of the outer product of the memory vector with itself, except that zeros are placed on the diagonal. Below is an example using the ASCII code for the letter Z:

$$
\begin{bmatrix} 0 \\ 1 \\ 0 \\ 1 \\ 1 \\ 0 \\ 1 \\ 0 \end{bmatrix} [0\ 1\ 0\ 1\ 1\ 0\ 1\ 0] =
\begin{bmatrix}
0 & 0 & 0 & 0 & 0 & 0 & 0 & 0 \\
0 & 0 & 0 & 1 & 1 & 0 & 1 & 0 \\
0 & 0 & 0 & 0 & 0 & 0 & 0 & 0 \\
0 & 1 & 0 & 0 & 1 & 0 & 1 & 0 \\
0 & 1 & 0 & 1 & 0 & 0 & 1 & 0 \\
0 & 0 & 0 & 0 & 0 & 0 & 0 & 0 \\
0 & 1 & 0 & 1 & 1 & 0 & 0 & 0 \\
0 & 0 & 0 & 0 & 0 & 0 & 0 & 0
\end{bmatrix}
$$

Hopfield (1982, 1984) has shown that the storage matrix must be symmetric, $W_{ij} = W_{ji}$, that W_{ii} must equal zero, and the matrix must be *dilute*. That is, there must be a smaller number of ones than zeros. in the matrix. Notice the lines formed in the matrix. They show at a glance the symmetric nature of the matrix.

This storage matrix can produce the correct memory state. That is, an eight-dimensional binary vector with bit errors, when multiplied by this storage matrix to give the inner product, will generate the correct memory state. The number of bit errors cannot be too great, but a partial memory will certainly work to give the complete memory state, for example:

$$\begin{bmatrix} 0 & 0 & 0 & 0 & 0 & 0 & 0 & 0 \\ 0 & 0 & 0 & 1 & 1 & 0 & 1 & 0 \\ 0 & 0 & 0 & 0 & 0 & 0 & 0 & 0 \\ 0 & 1 & 0 & 0 & 1 & 0 & 1 & 0 \\ 0 & 1 & 0 & 1 & 0 & 0 & 1 & 0 \\ 0 & 0 & 0 & 0 & 0 & 0 & 0 & 0 \\ 0 & 1 & 0 & 1 & 1 & 0 & 0 & 0 \\ 0 & 0 & 0 & 0 & 0 & 0 & 0 & 0 \end{bmatrix} \begin{bmatrix} 0 \\ 1 \\ 0 \\ 0 \\ 1 \\ 0 \\ 0 \\ 1 \end{bmatrix} = \begin{bmatrix} 0 \\ 1 \\ 0 \\ 1 \\ 1 \\ 0 \\ 1 \\ 0 \end{bmatrix}$$

The number of bit errors is called the *Hamming distance*. Given the two vectors

$$v = [0\ 1\ 0\ 0\ 1\ 0\ 0\ 1]$$

$$u = [0\ 1\ 0\ 1\ 1\ 0\ 1\ 0]$$

the Hamming distance is 3. Only vectors of equal dimensionality can be compared.

Hopfield (1982, 1984) has shown that if the matrix is symmetric and dilute with $W_{ii} = 0$, and if we define the dimension of the matrix as n, then m memories can be stored, where $m = 0.15n$. Table 5-1 is a list of

Table 5-1

Memories	Neurons (Matrix Size)
1	8
2	16
3	24
4	32
5	40

the number of memories for a given matrix size. Notice in the table that the actual number of memories has been rounded down to the nearest whole number. It doesn't make sense to store a fraction of a memory state. All of this can be expressed more formally to assist in writing code for a digital computer simulation.

The Hebb rule is used to determine the values of the W matrix. This is a vector outer product rule, as given below:

$$W_{ij} = \begin{cases} 1 & \text{if} \quad \sum_{i=1}^{R} v_i^s u_j^s > 0 \\ 0 & \text{otherwise} \end{cases}$$

This states that the element W_{ij} is found by summing the outer product of the input vector element j and the output vector element i. It is a simple outer product of these vectors. The W_{ij} element is then found by summing the W matrices. In other words, the outer product of the input vector u^s and the output vector v^s results in a matrix W^s. The elements of the final W matrix are found by summing the W^s matrices:

$$W = \sum_{s=1}^{\text{all states}} W^s$$

The sum is over all memory states. Each memory produces one matrix. The total memory matrix is the sum of these matrices. Hopfield (1982, 1984) has shown that if $W_{ii} = 0$ and $W_{ij} = W_{ji}$ then stable states exist, and the network will not oscillate chaotically.

Given the W matrix, the matrix vector product (the inner product) of this W with u^s (the input vector state) will result in the output vector v^s:

$$v^s = W u^s \qquad \text{(state } s\text{)}$$

The elements of this vector v^s are given by the following:

$$v_i^s = \sum_{j=1}^{N} W_{ij} u_j^s$$

This says that output vector element v_i is given by the sum of the products of elements $W_{ij} u_j$ summed over all j, for example:

$$W = \begin{bmatrix} W_{11} & W_{12} & W_{13} \\ W_{21} & W_{22} & W_{23} \\ W_{31} & W_{32} & W_{33} \end{bmatrix}$$

$$u_s = [u_1^s \quad u_2^s \quad u_3^s]$$

Then

$$v_s = \begin{bmatrix} W_{11} & W_{12} & W_{13} \\ W_{21} & W_{22} & W_{23} \\ W_{31} & W_{32} & W_{33} \end{bmatrix} \begin{bmatrix} u_1^s \\ u_2^s \\ u_3^s \end{bmatrix} = \begin{bmatrix} v_1^s \\ v_2^s \\ v_3^s \end{bmatrix}$$

$$v_1^s = W_{11}u_1^s + W_{12}u_2^s + W_{13}u_3^s$$

$$v_2^s = W_{21}u_1^s + W_{22}u_2^s + W_{23}u_3^s$$

$$v_3^s = W_{31}u_1^s + W_{32}u_2^s + W_{33}u_3^s$$

This output vector should include terms for the information input, bias, and noise. If these terms are added together to give one term I_i, then we get this:

$$v_i = \sum_j W_{ij}u_j + I_i$$

The actual vector is given by

$$v_i = \begin{cases} 1 & \text{if} \quad \sum_j W_{ij}u_j + I_i > I_i \\ 0 & \text{otherwise} \end{cases}$$

where I_t is threshold (see Fig. 5-5) and W_{ij} is the conductance of the connection between threshold logic units i and j. In other words, W_{ij} is the synaptic strength.

The energy corresponding to the stable states as given by Hopfield (1982, 1984) is

$$E = -\frac{1}{2}\sum_i \sum_j W_{ij}v_iv_j$$

where $W_{ij} = W_{ji}$ and $W_{ii} = 0$.

Goles and Vichniac (1986) write this equation in a form that clearly shows how to calculate the energy function:

$$E = -\frac{1}{2} \sum_{i=1}^{N} v_i^{t+1} \sum_{j=1}^{N} W_{ij} v_j^{t}$$

CONTENT-ADDRESSABLE MEMORIES AND ENERGY CALCULATIONS

When the inner product between a vector and a matrix is found, a vector is generated. These concepts were discussed earlier and will be reiterated here. If the starting vector doesn't differ too much from the stored memory state in the matrix, then the resulting vector is the correct memory state. This has obvious applications for content-addressable memory.

Some examples will make this more clear. Figure 5-6a shows a partial memory state for the complete memory state of Fig. 5-6b. If the partial memory state is digitized and operated on by an appropriate memory matrix, then the correct memory is generated or recalled.

A

B

Fig. 5-6.   Part A shows a partial memory state, B shows a complete memory state.

As another example, Fig. 5-7a shows a partial spectral pattern. In this partial memory state, no fine structure is observed. But when this spectrum is digitized and operated on by the appropriate storage matrix, then the spectrum of Figure 5-7b is recalled or generated.

This memory recall often happens to people, also. You see a person in the distance with lime-green socks, but other details are not clear. Then you recall that your friend has lime-green socks. This produces, in your mind, the entire picture of your friend.

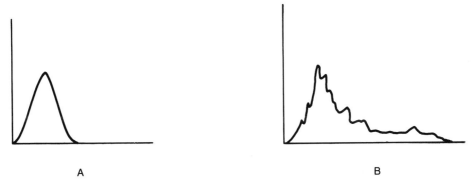

A   B

Fig. 5-7. Part A shows a partial spectral pattern, B shows a complete spectral pattern.

Further examples of content-addressable memory are easy to conceive. Looking in a toolbox, you might see only five percent of a wrench handle because the rest of the wrench is hidden by other tools. But this is enough of a pattern for you to recognize it as the wrench you need.

The first step in an algorithm for content-addressing is to find the inner product of a vector and a matrix:

$$v^s = Tu^s$$

The connection strength matrix T, is given by the following:

$$T = \begin{bmatrix} T_{11} & T_{12} & \cdots & T_{1N} \\ T_{21} & T_{22} & \cdots & T_2N \\ \vdots & & & \\ T_{i1} & T_{i2} & \cdots & T_{iN} \end{bmatrix}$$

It was shown earlier that the elements of the resulting vector from the inner product of T with vector u is as follows:

$$v_i^s = \sum_{j=1}^{N} T_{ij}u_j^s$$

It was further shown that the diagonal elements of the T matrix must be zero and the matrix should be symmetric and dilute. *Symmetric* means the following is true:

$$T_{ij} = T_{ji}$$

a dilute matrix means that the matrix contains more zeros than ones.

The program NEURON4P implements these ideas and calculates the inner product of a vector with a matrix. This simple program is not too useful by itself, but is used to build other programs. A simplified flow diagram of the program logic is shown in Fig. 5-8.

NEURON4P

```
10  CLS
20  INPUT "INPUT RANDOM SEED ";SEED
30  RANDOMIZE SEED
40  INPUT "ENTER THE NUMBER OF NEURONS (100 MAXIMUM) ";N
50  INPUT "INPUT THE THRESHOLD VALUE (0 TO 2 ARE REASONABLE
    VALUES) ";IO
60  INPUT "ENTER THE VALUE OF THE INFORMATION (0 TO 1 IS A
    GOOD VALUE)";INFO
70  INPUT "DO YOU WANT TO ENTER THE INPUT VECTOR YOURSELF
    (1/YES 0/NO)? ";VECTOR
80  INPUT "DO YOU WANT TO INPUT THE T MATRIX (1/Y 0/NO) ";
    MATRIX
90  DIM T(100,100),V(100),U(100)
100 REM FILL T(I,J) MATRIX
110 IF MATRIX=0 THEN 190
120 FOR I=1 TO N
130 FOR J=1 TO N
140 PRINT "T(";I;",";J;") "
150 INPUT T(I,J)
160 NEXT J
170 NEXT I
180 GOTO 360 : 'FILL INPUT VECTOR
190 FOR I=1 TO N
200 FOR J=I TO N
210 R=RND(1)
220 IF R<.75 THEN R=0 ELSE R=+1: REM DILUTE MATRIX
230 T(I,J)=R
240 NEXT J
250 LPRINT
260 NEXT I
270 FOR I=1 TO N
280 FOR J=1 TO N
290 IF I=J THEN T(I,J)=0
```

```
300 T(J,I)=T(I,J)
310 LPRINT T(I,J);
320 NEXT J
330 LPRINT
340 NEXT I
350 LPRINT:LPRINT:LPRINT
360 REM FILL INPUT VECTOR U
370 IF VECTOR=0 THEN 430
380 FOR I=1 TO N
390 PRINT "INPUT U(";I;")"
400 INPUT U(I)
410 NEXT I
420 GOTO 470 : 'BEGIN CALCULATIONS OF OUTPUT VECTOR
430 FOR I=1 TO N
440 GOSUB 670
450 U(I)=R
460 NEXT I
470 REM BEGIN CALCULATION
480 FOR I=1 TO N
490 FOR J=1 TO N
500 SIGMA=T(I,J)*U(J)+SIGMA
510 NEXT J
520 SIGMA=SIGMA+INFO
530 IF SIGMA > IO THEN SIGMA=1 ELSE SIGMA=0
540 V(I)=SIGMA
550 SIGMA=0
560 NEXT I
570 FOR I=1 TO N
580 LPRINT U(I);
590 NEXT I
600 LPRINT:LPRINT
610 FOR I=1 TO N
620 LPRINT V(I);
630 NEXT I
640 LPRINT:LPRINT
650 LPRINT:LPRINT
660 GOTO 360
670 R=RND(1)
680 IF R<.5 THEN R=0 ELSE R=+1
690 RETURN
```

The brief flowchart in Fig. 5-8 shows that after the T matrix is filled, then the input vector, u, is filled and the inner product is found between the T matrix and vector u. After vector v is computed, vectors u and v are printed to the line printer. Then a new u vector is selected and the process starts over again.

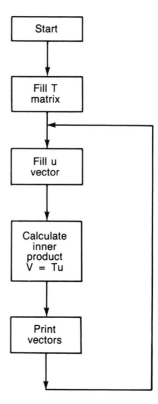

Fig. 5-8.   General flow chart for program
NEURONXP.

Now let's examine the program in more detail. In lines 10 and 20 a randomized seed is entered. In most digital computers, the random number generator needs a seed. Often the seed number can be generated by the timer. If the operator has control of the seed, then the same random number sequence is always generated. This is convenient for testing and developing programs.

In line 40 the number of neurons, N, is entered. A maximum has been set at 100 in the DIM statement of line 90. This can be changed as desired by the user.

Continuing with the theory, recall from Fig. 5-5 that the threshold value, I_t, is defined from the following equation:

$$V_i = \begin{cases} 1 & \text{if } \sum_j T_{ij}u_j + I_{info} > I_t \\ 0 & \text{otherwise} \end{cases}$$

The output of the ith neuron is logic 1 if the sum of the products of the elements $T_{ij}u_j$ and I_{info} is greater than threshold. Otherwise the output is logic 0.

The threshold is input in line 50 as the variable IO. For small networks, it is sufficient to choose zero or one for the threshold. This is equivalent to shifting the curve of Fig. 5-7 to the left or right. The information input to the neuron is entered in line 60. This information to the neuron can be thought of as a bias from another signal source and/or noise. It is simply called INFO in the program. By experimenting with the INFO and IO variables, you can get very different results.

Lines 70 and 80 ask if the user would like to enter the input vector and synapse matrix from the keyboard. If not, the program selects a random binary vector for input and also select a random, dilute, symmetric matrix with $T_{ii} = 0$.

Line 110 is another decision point, this time for either an operator-entered T matrix or a machine-generated matrix. Lines 120-170 allow the user to enter the matrix from the keyboard, while lines 190-350 are for a machine-generated matrix.

A random variable R between zero and one is selected in line 210. Line 220 dilutes the selection. If $R < 0.75$, then $R = 0$, else $R = 1$. The T_{ij} element being addressed is set equal to this R value in line 230. At the completion of line 260, the matrix is filled and dilute. Line 270 begins a routine to diagonalize the matrix, and in line 290, T_{ii} is set to zero. Line 300 symmetrizes the matrix with $T_{ij} = T_{ji}$. Finally, in line 310 the matrix is printed out to the line printer.

In line 360, the process to fill the input vector, u, is started. If the operator chooses to enter the vector from the keyboard, then lines 380–410 are executed; otherwise the machine selects a random binary vector in lines 430–460. This routine calls a subroutine, lines 670–690, to select a random number and decide if the vector element should be a zero or one.

In line 470, the calculation of the inner product of the input vector and the T matrix is started. In line 500, the variable named SIGMA is assigned to the result of this calculation. To this SIGMA value is added the INFO value in line 520. Finally, in line 530, SIGMA is compared with the threshold value IO and set equal to zero or one, depending on results. This final result becomes the output vector element v_i. After the v vector is filled by calculations, the u and v vectors are sent to the line printer in lines 580 and 620. In line 660, the process continues by selecting a new input vector u. The program ends only when the user presses CTRL-BREAK.

Now that this program has been discussed in detail, we can move on to other programs. Table 5-2 summarizes the programs and their differences. These are explained in a little more detail below.

Table 5-2

Program	Comments
NEURON4P	Basic program to find inner product of a vector and a matrix. $T_{ij} = T_{ji}$, $T_{ii} = 0$
NEURON5P	Same as NEURON4P but matrix is random and a little more dilute. $T_{ij} = T_{ji}$
NEURON6P	Iterated version of NEURON5P. Runs for eight iterations then prints resulting vector.
NEURON8P	Includes energy calculations betweeen each iteration.
HEBB2P	Includes one memory vector.
HEBB3P	Includes m memory vectors.

The next program to be examined is NEURON5P. This program shows the effects of a random dilute matrix. In line 210 the random element is chosen, and in line 220 the matrix is diluted. In this program the matrix is more dilute than in NEURON4P. In this program if $R < 0.85$ then $R = 0$, else R is set equal to 1. This results in less than 15% of the elements being set to 1. In NEURON4P the dilution was 25%. The rest of this program is similar to NEURON4P.

NEURON5P

```
10 CLS
20 INPUT "INPUT RANDOM SEED ";SEED
30 RANDOMIZE SEED
40 INPUT "ENTER THE NUMBER OF NEURONS (100 MAXIMUM) ";N
50 INPUT "INPUT THE THRESHOLD VALUE (0 TO 2 ARE REASONABLE
   VALUES) ";IO
60 INPUT "ENTER THE VALUE OF THE INFORMATION (0 TO 1 IS A
GOOD VALUE ) ";INFO
70 INPUT "DO YOU WANT TO ENTER THE INPUT VECTOR YOURSELF
   (1/YES 0/NO)? ";VECTOR
80 INPUT "DO YOU WANT TO INPUT THE T MATRIX (1/Y 0/NO) ";
   MATRIX
90 DIM T(100,100),V(100),U(100)
100 REM FILL T(I,J) MATRIX
110 IF MATRIX=0 THEN 190
120 FOR I=1 TO N
```

```
130 FOR J=1 TO N
140 PRINT "T(";I;",";J;") "
150 INPUT T(I,J)
160 NEXT J
170 NEXT I
180 GOTO 360 : 'FILL INPUT VECTOR
190 FOR I=1 TO N
200 FOR J=1 TO N
210 R=RND(1)
220 IF R<.85 THEN R=0 ELSE R=+1: REM DILUTE MATRIX
230 T(I,J)=R
240 NEXT J
250 LPRINT
260 NEXT I
270 FOR I=1 TO N
280 FOR J=1 TO N
290 IF I=J THEN T(I,J)=0
300 REM     T(J,I)=T(I,J)
310 LPRINT T(I,J);
320 NEXT J
330 LPRINT
340 NEXT I
350 LPRINT:LPRINT:LPRINT
360 REM FILL INPUT VECTOR U
370 IF VECTOR=0 THEN 430
380 FOR I=1 TO N
390 PRINT "INPUT U(";I;")"
400 INPUT U(I)
410 NEXT I
420 GOTO 470 : 'BEGIN CALCULATIONS OF OUTPUT VECTOR
430 FOR I=1 TO N
440 GOSUB 670
450 U(I)=R
460 NEXT I
470 REM BEGIN CALCULATION
480 FOR I=1 TO N
490 FOR J=1 TO N
500 SIGMA=T(I,J)*U(J)+SIGMA
510 NEXT J
520 SIGMA=SIGMA+INFO
530 IF SIGMA > IO THEN SIGMA=1 ELSE SIGMA=0
540 V(I)=SIGMA
550 SIGMA=0
560 NEXT I
570 FOR I=1 TO N
580 LPRINT U(I);
590 NEXT I
600 LPRINT:LPRINT
```

```
610 FOR I=1 TO N
620 LPRINT V(I);
630 NEXT I
640 LPRINT:LPRINT
650 LPRINT:LPRINT
660 GOTO 360
670 R=RND(1)
680 IF R<.5 THEN R=0 ELSE R=+1
690 RETURN
```

A run of this program produces some stable states but it also produces many superfluous states. Stable states can be thought of as low points on an energy hyperplane in an n-dimensional hypercube. This is explained in more detail later in this chapter. It is interesting to note that a run of program NEURON5P results in some spurious states. There are two problems. One is that the matrix is nonsymmetric. The second is that *iterative dynamics* have not been implemented in this program.

Iterative dynamics means that the results from one vector-matrix product should be sent back into the network and operated on again by the same matrix. Only by iterative operation can these massively parallel networks compute stable states. As pointed out earlier, the neurons or threshold logic processors are connected to each other through a synapse or conductance matrix. The output of a processor may be connected to the input of several other processors. Each threshold logic device sums the inputs it receives, so the signal travels around this feedback loop, but not to itself, many times in a second before the stable state is reached. Iterative dynamics are introduced in the next program.

In NEURON6P, a new variable, ITERATE, is introduced. This variable is a counter in a loop that starts in line 480. The output vector from the inner product of the input vector and the matrix is set equal to a new input vector in line 640. The inner product of this vector and the matrix is then found and the process repeated eight times. (Later you will see that eight times is a few too many. Only three or four times are needed.)

NEURON6P

```
10 CLS
20 INPUT "INPUT RANDOM SEED ";SEED
30 RANDOMIZE SEED
40 INPUT "ENTER THE NUMBER OF NEURONS (100 MAXIMUM) ";N
50 INPUT "INPUT THE THRESHOLD VALUE (0 TO 2 ARE REASONABLE
   VALUES) ";IO
60 INPUT "ENTER THE VALUE OF THE INFORMATION (0 TO 1 IS A
   GOOD VALUE ) ";INFO
```

```
70 INPUT "DO YOU WANT TO ENTER THE INPUT VECTOR YOURSELF
   (1/YES 0/NO)? ";VECTOR
80 INPUT "DO YOU WANT TO INPUT THE T MATRIX (1/Y 0/NO) ";
   MATRIX
90 DIM T(100,100),V(100),U(100)
100 REM FILL T(I,J) MATRIX
110 IF MATRIX=0 THEN 190
120 FOR I=1 TO N
130 FOR J=1 TO N
140 PRINT "T(";I;",";J;") "
150 INPUT T(I,J)
160 NEXT J
170 NEXT I
180 GOTO 360 : 'FILL INPUT VECTOR
190 FOR I=1 TO N
200 FOR J=1 TO N
210 R=RND(1)
220 IF R<.8 THEN R=0 ELSE R=+1: REM DILUTE MATRIX
230 T(I,J)=R
240 NEXT J
250 LPRINT
260 NEXT I
270 FOR I=1 TO N
280 FOR J=1 TO N
290 IF I=J THEN T(I,J)=0
300           T(J,I)=T(I,J)
310 LPRINT T(I,J);
320 NEXT J
330 LPRINT
340 NEXT I
350 LPRINT:LPRINT:LPRINT
360 REM FILL INPUT VECTOR U
370 IF VECTOR=0 THEN 430
380 FOR I=1 TO N
390 PRINT "INPUT U(";I;")"
400 INPUT U(I)
410 NEXT I
420 GOTO 470 : 'BEGIN CALCULATIONS OF OUTPUT VECTOR
430 FOR I=1 TO N
440 GOSUB 720
450 U(I)=R
460 NEXT I
470 REM BEGIN CALCULATION
480 FOR ITERATE=1 TO 8: REM THIS ALLOWS THE OUTPUT VECTOR
    TO BE FEED BACK
490 FOR I=1 TO N
500 FOR J=1 TO N
510 SIGMA=T(I,J)*U(J)+SIGMA
```

```
520 NEXT J
530 SIGMA=SIGMA+INFO
540 IF SIGMA > IO THEN SIGMA=1 ELSE SIGMA=0
550 V(I)=SIGMA
560 SIGMA=0
570 NEXT I
580 IF ITERATE=1 THEN 590 ELSE 630
590 FOR I=1 TO N
600 LPRINT U(I);
610 NEXT I
620 LPRINT
630 FOR I=1 TO N
640 U(I)=V(I): REM FOR FEEDBACK
650 NEXT I
660 NEXT ITERATE
670 FOR I=1 TO N
680 LPRINT V(I);
690 NEXT I
700 LPRINT:LPRINT:LPRINT:LPRINT
710 GOTO 360
720 R=RND(1)
730 IF R<.5 THEN R=0 ELSE R=+1
740 RETURN
```

This program prints the matrix. Then it prints the initial input vector followed by the eighth iterated resulting vector as the output. A new initial vector is then selected and the process started over again. A run of this program is shown in Fig. 5-9. Notice that there still appears to be more than one stable state. Table 5-1 shows that an eight-neuron circuit can have only one stable state. This can be explained from an energy consideration.

By the iterative dynamics, the algorithm computes the minimal states in an n-dimensional hyperspace. By definition, there is more than one minimum on this hypersurface. In fact, all the corners of the hypercube are stable states. These do not all have the same degree of stability, and may only be metastable states. These minima are strange attractors.

A convenient way of thinking about the energy surface is as a sheet of rubber being pulled in many directions, from many points on its surface. This results in a surface with hills, valleys and wells. If a small marble is dropped on this surface it will be attracted to the nearest lowest point. This point might not be the lowest point in the entire hypersurface; it is just a local minimum or attractor. If the marble is kicked around hard enough, it will jump out of this local minimum and settle to the next.

MATRIX

```
0  0  0  0  0  0  0  1
0  0  1  0  0  0  0  0
0  1  0  0  0  0  0  0
0  0  0  0  0  1  0  0
0  0  0  0  0  0  0  0
0  0  0  1  0  0  0  0
0  0  0  0  0  0  0  1
1  0  0  0  0  0  1  0
```

Input 1	0	0	0	1	0	1	1	0		**Input 12**	0	1	1	1	1	1	1	0
Output 1	1	0	0	1	0	1	1	0		**Output 12**	1	1	1	1	0	1	1	0
Input 2	0	0	0	0	1	1	0	1		**Input 13**	1	1	0	0	0	0	0	1
Output 2	0	0	0	0	0	1	0	1		**Output 13**	1	1	0	0	0	0	1	1
Input 3	0	0	1	0	0	1	0	0		**Input 14**	1	0	1	0	0	0	0	0
Output 3	0	0	1	0	0	1	0	0		**Output 14**	1	0	1	0	0	0	1	0
Input 4	1	0	1	0	1	1	0	1		**Input 15**	1	1	0	1	0	0	1	1
Output 4	1	0	1	0	0	1	1	1		**Output 15**	1	1	0	1	0	0	1	1
Input 5	0	1	1	0	0	0	1	1		**Input 16**	1	0	1	0	0	0	1	1
Output 5	1	1	1	0	0	0	1	1		**Output 16**	1	0	1	0	0	0	1	1
Input 6	1	1	1	0	0	1	1	0		**Input 17**	1	1	1	1	1	1	1	0
Output 6	1	1	1	0	0	1	1	0		**Output 17**	1	1	1	1	0	1	1	0
Input 7	0	0	1	1	0	1	0	1		**Input 18**	0	1	0	0	0	0	0	1
Output 7	0	0	1	1	0	1	0	1		**Output 18**	0	1	0	0	0	0	0	1
Input 8	0	1	0	1	1	1	1	1		**Input 19**	1	0	1	0	1	0	0	0
Output 8	1	1	0	1	0	1	1	1		**Output 19**	1	0	1	0	0	0	1	0
Input 9	0	1	0	1	0	0	0	1		**Input 20**	0	0	1	1	0	0	1	1
Output 9	0	1	0	1	0	0	0	1		**Output 20**	1	0	1	1	0	0	1	1
Input 10	0	1	0	0	0	1	1	1		**Input 21**	1	1	0	1	0	0	1	1
Output 10	1	1	0	0	0	1	1	1		**Output 21**	1	1	0	1	0	0	1	1
Input 11	1	0	1	1	0	1	1	0		**Input 22**	1	0	1	0	0	1	0	1
Output 11	1	0	1	1	0	1	1	0		**Output 22**	1	0	1	0	0	1	1	1

Fig. 5-9.   Example run of program NEURON6P. Seed 72873, threshold 1, information 1.

Repeating this process will result in the marble settling in the deepest basin of attraction. This is the most stable memory state of the network.

The computed result shows several minima. In *Artificial Neural Networks* (TAB BOOKS Inc., 1988) I show how to build actual electronic neural networks. Because of Johnson noise and other component noise, I show that only one stable state results for an eight-neuron circuit. This noise is the analogous effect of kicking the marble around until it settles in the deepest basin of attraction.

The next program calculates the energy after each iteration. Goles and Vichniac (1986) give the energy calculation for a Hopfield (1982, 1984) model as follows:

$$E(t) = -\frac{1}{2}\sum_i V_i^{t+1}\sum_j T_{ij}V_j^t$$

This algorithm says that the inner product of vector *v* at time t with connection strength matrix *T* is multiplied with the vector *v* at time $t + 1$. Let time t count the number of sweeps through the network; and $t + 1$ is the next sweep.

This second multiplication is a dot product of two vectors and results in a scalar. This scalar value is proportional to the energy. From this dynamical equation it is clear that the energy reaches a minimum when the following is true:

$$\sum_i V_i^{t+1} = \sum_j T_{ij}V_j^t$$

This is clearly seen in a computer simulation. After a few iterations, usually four or less, the energy settles to a stable point. Now let's examine the program, NEURON8P.

NEURON8P

```
10 CLS
20 INPUT "INPUT RANDOM SEED ";SEED
30 RANDOMIZE SEED
40 INPUT "ENTER THE NUMBER OF NEURONS (100 MAXIMUM) ";N
50 INPUT "INPUT THRESHOLD VALUE (0 TO 2 ARE REASONABLE
   VALUES) ";IO
60 INPUT "ENTER THE VALUE OF THE INFORMATION (0 TO 1 IS
   A GOOD VALUE ) ";INFO
70 INPUT "DO YOU WANT TO ENTER THE INPUT VECTOR YOURSELF
   (1/YES 0/NO)? ";VECTOR
```

```
80 INPUT "DO YOU WANT TO INPUT THE T MATRIX (1/Y 0/NO) ";
   MATRIX
90 DIM T(100,100),V(100),U(100)
100 REM FILL T(I,J) MATRIX
110 IF MATRIX=0 THEN 190
120 FOR I=1 TO N
130 FOR J=1 TO N
140 PRINT "T(";I;",";J;") "
150 INPUT T(I,J)
160 NEXT J
170 NEXT I
180 GOTO 360 : 'FILL INPUT VECTOR
190 FOR I=1 TO N
200 FOR J=1 TO N
210 R=RND(1)
220 IF R<.8 THEN R=0 ELSE R=+1: REM DILUTE MATRIX
230 T(I,J)=R
240 NEXT J
250 LPRINT
260 NEXT I
270 FOR I=1 TO N
280 FOR J=1 TO N
290 IF I=J THEN T(I,J)=0
300            T(J,I)=T(I,J)
310 LPRINT T(I,J);
320 NEXT J
330 LPRINT
340 NEXT I
350 LPRINT:LPRINT:LPRINT
360 REM FILL INPUT VECTOR U
370 IF VECTOR=0 THEN 430
380 FOR I=1 TO N
390 PRINT "INPUT U(";I;")"
400 INPUT U(I)
410 NEXT I
420 GOTO 470 : 'BEGIN CALCULATIONS OF OUTPUT VECTOR
430 FOR I=1 TO N
440 GOSUB 790
450 U(I)=R
460 NEXT I
470 REM BEGIN CALCULATION
480 FOR ITERATE=1 TO 8:REM THIS ALLOWS THE OUTPUT VECTOR
    TO BE FEED BACK
490 FOR I=1 TO N
```

```
500 FOR J=1 TO N
510 SIGMA=T(I,J)*U(J)+SIGMA
520 NEXT J
530 SIGMA=SIGMA+INFO
540 IF SIGMA > IO THEN SIGMA=1 ELSE SIGMA=0
550 V(I)=SIGMA
560 SIGMA=0
570 NEXT I
580 FOR I=1 TO N
590 LPRINT U(I);
600 NEXT I
610 LPRINT
620 FOR I=1 TO N
630 LPRINT V(I);
640 NEXT I
650 LPRINT
660 ENERGY=0
670 FOR I=1 TO N : REM ENERGY CALCULATION
680 ENERGY=ENERGY+(U(I)*V(I))
690 NEXT I
700 ENERGY=-.5*ENERGY
710 LPRINT "     ENERGY ";ENERGY:PRINT
720 FOR I=1 TO N
730 U(I)=V(I): REM FOR FEEDBACK
740 NEXT I
750 ENERGY=0
760 NEXT ITERATE
770 LPRINT:LPRINT:LPRINT:LPRINT
780 GOTO 360
790 R=RND(1)
800 IF R<.5 THEN R=0 ELSE R=+1
810 RETURN
```

The program NEURON8P introduces the new variable ENERGY. The energy calculation takes place within the ITERATE loop. The energy is set equal to zero in line 660, and calculations take place in a loop starting at 670 and ending at 690. The final energy is printed out in line 700. In line 750, the energy is set equal to zero again and the next iteration begins.

Figure 5-10 shows a run of this program. After the matrix is printed, the random binary vector is printed. Then the resulting inner product vector and energy are printed. The program then begins the next iteration by feeding the resulting vector back into the matrix calculation.

MATRIX

```
O  O  O  O  1  O  O  O  O  O  1  1  O  O  O  O
O  O  O  O  O  O  O  O  O  O  O  1  O  O  O  O
O  O  O  O  O  O  1  O  O  O  O  O  1  O  O  1
O  O  O  O  O  O  O  O  1  O  1  O  O  O  O  O
1  O  O  O  O  1  O  O  O  1  O  O  O  O  1  1
O  O  O  O  1  O  O  O  O  O  O  O  O  O  O  O
O  O  1  O  O  O  O  O  O  O  O  1  O  O  O  O
O  O  O  O  O  O  O  O  1  O  1  1  O  1  O  O
O  O  O  1  O  O  O  1  O  O  O  1  O  1  O  1
O  O  O  O  1  O  O  O  O  O  1  1  O  O  O  O
1  O  O  1  O  O  O  1  O  1  O  O  O  O  O  1
1  1  O  O  O  O  1  1  1  1  O  O  O  O  O  O
O  O  1  O  O  O  O  O  O  O  O  O  O  O  O  1
O  O  O  O  O  O  O  1  1  O  O  O  O  O  O  O
O  O  O  O  1  O  O  O  O  O  O  O  O  O  O  O
O  O  1  O  1  O  O  O  1  O  1  O  1  O  O  O
```

Input 1	1	O	1	O	1	O	1	1	1	1	O	1	1	O	O	O
Output 1	1	O	1	O	1	O	1	1	1	1	1	1	O	1	O	1
		ENERGY	−4													
Input 2	1	O	1	O	1	O	1	1	1	1	1	1	O	1	O	1
Output 2	1	O	1	1	1	O	1	1	1	1	1	1	1	O	1
		ENERGY	−5.5													
Input 3	1	O	1	1	1	O	1	1	1	1	1	1	1	O	1
Output 3	1	O	1	1	1	O	1	1	1	1	1	1	1	O	1
		ENERGY	−6.5													
Input 4	1	O	1	1	1	O	1	1	1	1	1	1	1	O	1
Output 4	1	O	1	1	1	O	1	1	1	1	1	1	1	O	1
		ENERGY	−6.5													
Input 5	1	O	1	1	1	O	1	1	1	1	1	1	1	O	1
Output 5	1	O	1	1	1	O	1	1	1	1	1	1	1	O	1
		ENERGY	−6.5													
Input 6	1	O	1	1	1	O	1	1	1	1	1	1	1	O	1
Output 6	1	O	1	1	1	O	1	1	1	1	1	1	1	O	1
		ENERGY	−6.5													
Input 7	1	O	1	1	1	O	1	1	1	1	1	1	1	O	1
Output 7	1	O	1	1	1	O	1	1	1	1	1	1	1	O	1
		ENERGY	−6.5													
Input 8	1	O	1	1	1	O	1	1	1	1	1	1	1	O	1
Output 8	1	O	1	1	1	O	1	1	1	1	1	1	1	O	1
		ENERGY	−6.5													

Fig. 5-10.   Example run of program NEURON8P. Seed 0, threshold 1, information 1.

Input 9	1	O	1	1	1	O	1	O	1	1	O	1	O	1	1	1
Output 9	1	O	1	O	1	O	1	1	1	1	1	1	1	O	O	1

ENERGY −4

Input 10	1	O	1	O	1	O	1	1	1	1	1	1	1	O	O	1
Output 10	1	O	1	1	1	O	1	1	1	1	1	1	1	1	O	1

ENERGY −5.5

Input 11	1	O	1	1	1	O	1	1	1	1	1	1	1	1	O	1
Output 11	1	O	1	1	1	O	1	1	1	1	1	1	1	1	O	1

ENERGY −6.5

Input 12	1	O	1	1	1	O	1	1	1	1	1	1	1	1	O	1
Output 12	1	O	1	1	1	O	1	1	1	1	1	1	1	1	O	1

ENERGY −6.5

Input 13	1	O	1	1	1	O	1	1	1	1	1	1	1	1	O	1
Output 13	1	O	1	1	1	O	1	1	1	1	1	1	1	1	O	1

ENERGY −6.5

Input 14	1	O	1	1	1	O	1	1	1	1	1	1	1	1	O	1
Output 14	1	O	1	1	1	O	1	1	1	1	1	1	1	1	O	1

ENERGY −6.5

Input 15	1	O	1	1	1	O	1	1	1	1	1	1	1	1	O	1
Output 15	1	O	1	1	1	O	1	1	1	1	1	1	1	1	O	1

ENERGY −6.5

Input 16	1	O	1	1	1	O	1	1	1	1	1	1	1	1	O	1
Output 16	1	O	1	1	1	O	1	1	1	1	1	1	1	1	O	1

ENERGY −6.5

Input 17	1	1	O	1	1	1	O	O	1	1	O	O	O	1	O	O
Output 17	O	O	O	O	1	O	O	1	1	O	1	1	O	O	O	1

ENERGY −1

Input 18	O	O	O	O	1	O	O	1	1	O	1	1	O	O	O	1
Output 18	1	O	O	1	O	O	O	1	1	1	1	1	O	1	O	1

ENERGY −2.5

Input 19	1	O	O	1	O	O	O	1	1	1	1	1	O	1	O	1
Output 19	1	O	O	1	1	O	O	1	1	1	1	1	O	1	O	1

ENERGY −4.5

Input 20	1	O	O	1	1	O	O	1	1	1	1	1	O	1	O	1
Output 20	1	O	O	1	1	O	O	1	1	1	1	1	O	1	O	1

ENERGY −5

Input 21	1	O	O	1	1	O	O	1	1	1	1	1	O	1	O	1
Output 21	1	O	O	1	1	O	O	1	1	1	1	1	O	1	O	1

ENERGY −5

Input 22	1	O	O	1	1	O	O	1	1	1	1	1	O	1	O	1
Output 22	1	O	O	1	1	O	O	1	1	1	1	1	O	1	O	1

ENERGY −5

Input 23	1	O	O	1	1	O	O	1	1	1	1	1	O	1	O	1
Output 23	1	O	O	1	1	O	O	1	1	1	1	1	O	1	O	1

ENERGY −5

Input 24	1	O	O	1	1	O	O	1	1	1	1	1	O	1	O	1
Output 24	1	O	O	1	1	O	O	1	1	1	1	1	O	1	O	1

ENERGY −5

Fig. 5-10. (Continued)

Looking at Figure 5-10 shows that, in the case of the first random binary vector, the energy has a value of -4 after the first iteration and -5.5 after the second iteration, finally settling to a stable state at -6.5 energy units. The next two programs examine associative learning using the Hebb learning rule.

ASSOCIATIVE LEARNING

The program HEBB2P is a basic building unit for the Hebb learning rule. This rule was introduced earlier, but is reiterated here. The original Hebb learning rule (Hebb, 1949) was not sufficiently quantitative to build a good model. The rule in its original version states that if neuron *A* and neuron *B* are simultaneously excited, then the synaptic connection strength between them is increased.

HEBB2P

```
10 CLS
20 INPUT "INPUT RANDOM SEED ";SEED
30 RANDOMIZE SEED
40 INPUT "ENTER THE NUMBER OF NEURONS (100 MAXIMUM) ";N
50 INPUT "INPUT THE THRESHOLD VALUE (0 TO 2 ARE REASONABLE
   VALUES) ";IO
60 INPUT "ENTER THE VALUE OF THE INFORMATION (0 TO 1 IS A
   GOOD VALUE ) ";INFO
70 INPUT "DO YOU WANT TO ENTER THE INPUT VECTOR YOURSELF
   (1/YES 0/NO)? ";VECTOR
80 PRINT "BINARY MATRIX WITH Tii=0 AND Tij=Tji."
90 DIM T(100,100),V(100),U(100)
100 REM FILL T(I,J) MATRIX
110 PRINT:PRINT:PRINT
120 PRINT "INPUT THE MEMORY VECTOR FOR THE HEBB MATRIX"
130 FOR I=1 TO N
140 PRINT "V(";I;")"
150 INPUT V(I)
160 U(I)=V(I)
170 NEXT I
180 FOR I=1 TO N
190 FOR J=1 TO N
200 T(I,J)=V(I)*U(J)
210 IF I=J THEN T(I,J)=0
220 LPRINT T(I,J);
230 NEXT J
240 LPRINT
250 NEXT I
260 LPRINT:LPRINT:LPRINT
270 REM FILL INPUT VECTOR U
280 IF VECTOR=0 THEN 340
```

```
290 FOR I=1 TO N
300 PRINT "INPUT U(";I;")"
310 INPUT U(I)
320 NEXT I
330 GOTO 380 : 'BEGIN CALCULATIONS OF OUTPUT VECTOR
340 FOR I=1 TO N
350 GOSUB 630
360 U(I)=R
370 NEXT I
380 REM BEGIN CALCULATION
390 FOR ITERATE=1 TO 8: REM THIS ALLOWS THE OUTPUT VECTOR
    TO BE FEED BACK
400 FOR I=1 TO N
410 FOR J=1 TO N
420 SIGMA=T(I,J)*U(J)+SIGMA
430 NEXT J
440 SIGMA=SIGMA+INFO
450 IF SIGMA > IO THEN SIGMA=1 ELSE SIGMA=0
460 V(I)=SIGMA
470 SIGMA=0
480 NEXT I
490 IF ITERATE=1 THEN 500 ELSE 540
500 FOR I=1 TO N
510 LPRINT U(I);
520 NEXT I
530 LPRINT
540 FOR I=1 TO N
550 U(I)=V(I): REM FOR FEEDBACK
560 NEXT I
570 NEXT ITERATE
580 FOR I=1 TO N
590 LPRINT V(I);
600 NEXT I
610 LPRINT:LPRINT:LPRINT:LPRINT
620 GOTO 270
630 R=RND(1)
640 IF R<.5 THEN R=0 ELSE R=+1
650 RETURN
```

An excellent example of associative learning in humans is holding a red apple in front of a baby and repeatedly saying red. Synaptic connection strengths are increased when the appropriate neurons from the optic center are simultaneously activated with those from the auditory center for the sound of the word red. Another example is Pavlov's experiments, in which, after repeated trials, a dog learned to associate the sound of a bell with food.

This learning rule could be used to train a simple network such as that shown in Fig. 5-11. An input vector would be presented to both the

AUDITORY NEURONS

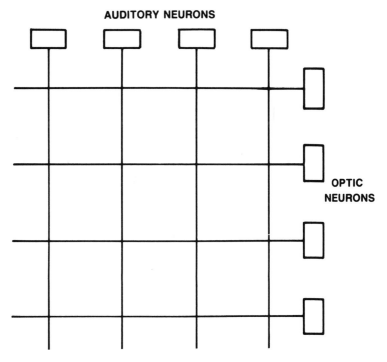

OPTIC
NEURONS

Fig. 5-11.   Simplified network of optic and auditory neurons.

auditory and optic neurons. The appropriate synaptic connections would then be strengthened.

In digital simulations of this model, the outer product of two vectors is found to produce a synaptic connection strength matrix. This is given symbolically as follows:

$$W = uv^t$$

For a one-dimensional vector of length four, we would get the matrix W:

$$u = \begin{bmatrix} 3 & 1 & 2 & 4 \end{bmatrix}$$

$$v = \begin{bmatrix} 0 & 1 & 1 & 6 \end{bmatrix}$$

$$w = \begin{bmatrix} 3 & 1 & 2 & 4 \end{bmatrix} \begin{bmatrix} 0 \\ 1 \\ 1 \\ 6 \end{bmatrix} = \begin{bmatrix} 0 & 0 & 0 & 0 \\ 3 & 1 & 2 & 4 \\ 3 & 1 & 2 & 4 \\ 18 & 6 & 12 & 24 \end{bmatrix}$$

Notice we used the transpose of the vector v. Symbolically, to find an element of the matrix, we write the following:

$$W_{ij} = u_i v_j^t$$

The connection strengths are the elements of the matrix. These represent the stored memory state or states. If the storage matrix is small, then only one memory state can be stored. For larger storage matrices, more than one memory can be stored. In that case, each memory state generates one matrix:

$$W^s = u^s (v^t)^s \quad \text{(state s)}$$

To store all the memories in one matrix, the matrices are summed over all states:

$$W = \sum_{\text{all states}} W^s$$

Table 15-1 summarizes the number of memories versus the number of neurons. In order to recover the memory state from the storage matrix, the inner product of a partial memory and the memory matrix is calculated.

Figure 5-12 is a simplified flow diagram of the program logic of HEBB2P. The first observation is that there is no END Statement. In order to end, the CTRL-BREAK key combination must be pressed. Another obvious observation is that only one memory state is stored in this matrix. This program is used to develop the next program, HEBB3P, which can store m memory states for N neurons.

HEBB3P

```
10 CLS
20 INPUT "INPUT RANDOM SEED ";SEED
30 RANDOMIZE SEED
40 INPUT "ENTER THE NUMBER OF NEURONS (100 MAXIMUM) ";N
50 INPUT "DO YOU WANT TO ENTER THE INPUT VECTOR YOURSELF
   (1/YES 0/NO)? ";VECTOR
60 DIM T(100,100),V(100),U(100)
70 REM FILL T(I,J) MATRIX
80 PRINT:PRINT:PRINT
90 INPUT "INPUT THE NUMBER OF MEMORY VECTORS (M=INT
   (.15*N) ";M
100 FOR MEMS=1 TO M
110 PRINT "INPUT THE MEMORY VECTOR ";MEMS;"FOR THE HEBB
    MATRIX."
```

```
120 FOR I=1 TO N
130 PRINT "V(";I;")"
140 INPUT V(I)
150 U(I)=V(I)
160 NEXT I
170 FOR I=1 TO N
180 FOR J=1 TO N
190 T(I,J)=T(I,J)+V(I)*U(J)
200 IF I=J THEN T(I,J)=0
210 IF T(I,J)>1 THEN T(I,J)=1
220 LPRINT T(I,J);
230 NEXT J
240 LPRINT
250 NEXT I
255 LPRINT:LPRINT:LPRINT:LPRINT:LPRINT
260 NEXT MEMS
270 LPRINT:LPRINT:LPRINT
280 REM FILL INPUT VECTOR U
290 IF VECTOR=0 THEN 350
300 FOR I=1 TO N
310 PRINT "INPUT U(";I;")"
320 INPUT U(I)
330 NEXT I
340 GOTO 390 : 'BEGIN CALCULATIONS OF OUTPUT VECTOR
350 FOR I=1 TO N
360 GOSUB 640
370 U(I)=R
380 NEXT I
390 REM BEGIN CALCULATION
400 FOR ITERATE=1 TO 8: REM THIS ALLOWS THE OUTPUT VECTOR
    TO BE FEED BACK
410 FOR I=1 TO N
420 FOR J=1 TO N
430 SIGMA=T(I,J)*U(J)+SIGMA
440 NEXT J
450 SIGMA=SIGMA
460 IF SIGMA > 0  THEN SIGMA=1 ELSE SIGMA=0
470 V(I)=SIGMA
480 SIGMA=0
490 NEXT I
500 IF ITERATE=1 THEN 510 ELSE 550
510 FOR I=1 TO N
520 LPRINT U(I);
530 NEXT I
540 LPRINT
550 FOR I=1 TO N
560 U(I)=V(I): REM FOR FEEDBACK
570 NEXT I
580 NEXT ITERATE
```

```
590 FOR I=1 TO N
600 LPRINT V(I);
610 NEXT I
620 LPRINT:LPRINT:LPRINT:LPRINT
630 GOTO 280
640 R=RND(1)
650 IF R<.5 THEN R=0 ELSE R=+1
660 RETURN
```

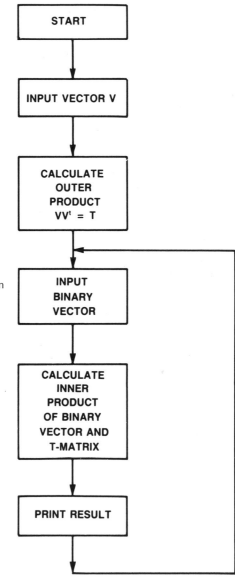

Fig. 5-12.   Flow chart for program HEBBXP.

Looking at the program line by line, you can see that the memory state is entered in lines 130 to 170. In line 160, the transpose is found by changing the label. Beginning in line 180, the *T* matrix is filled by finding the outer product of the desired storage vector and its transpose. Line 210 puts zeros on the diagonal of the matrix and line 220 prints the *T* matrix at the line printer.

The rest of the program is exactly like earlier programs. A random binary vector is filled, then the inner product of this vector with the *T* matrix is calculated. This calculation is iterated eight times, and the random vector and final product vector are printed at the line printer.

Program HEBB3P includes a routine to allow the user to choose the number of memory states to store. This program asks the user how many neurons are to be simulated. It then asks how many memory states are to be stored and reminds the user that the number of memory states stored is given by

$$m = \text{INT}(0.15N)$$

where *N* is the number of neurons. This equation for the number of memories is empirically derived. You can experiment with these programs and deduce more or less the same equation for number of memory states.

After the user enters the first memory vector, the first *T* matrix is found and printed at the line printer. The next memory vector is entered, and the new *T* matrix is printed. This new *T* matrix includes the sum of the *T* matrix for state one and state two. This continues through m memory states. Then a random binary vector is chosen or entered from the keyboard, and the inner product of this vector and the summed *T* matrix is found and printed at the line printer along with the random vector.

Figure 5-13 is a run of this program with 16 neurons and two memory states. The two states entered were as follows:

(1 0 0 0 1 0 0 0 1 0 0 0 1 0 0 0)

(0 1 1 0 0 0 1 1 0 0 0 0 0 1 1 0)

Notice that these states only come out from random vectors twice for the second vector and once for the first vector. In the other cases, the Hamming distance is too great to result in a correct memory state. The end result is a stable spurious state. You can see that this same stable spurious state arises many times, indicating that it is probably a deeper energy minima than the two stored states.

MEMORY STATE 1

```
O  O  O  O  1  O  O  O  1  O  O  O  1  O  O  O
O  O  O  O  O  O  O  O  O  O  O  O  O  O  O  O
O  O  O  O  O  O  O  O  O  O  O  O  O  O  O  O
O  O  O  O  O  O  O  O  O  O  O  O  O  O  O  O
1  O  O  O  O  O  O  O  O  1  O  O  O  1  O  O  O
O  O  O  O  O  O  O  O  O  O  O  O  O  O  O  O
O  O  O  O  O  O  O  O  O  O  O  O  O  O  O  O
O  O  O  O  O  O  O  O  O  O  O  O  O  O  O  O
1  O  O  O  1  O  O  O  O  O  O  O  O  1  O  O  O
O  O  O  O  O  O  O  O  O  O  O  O  O  O  O  O
O  O  O  O  O  O  O  O  O  O  O  O  O  O  O  O
O  O  O  O  O  O  O  O  O  O  O  O  O  O  O  O
1  O  O  O  1  O  O  O  1  O  O  O  O  O  O  O
O  O  O  O  O  O  O  O  O  O  O  O  O  O  O  O
O  O  O  O  O  O  O  O  O  O  O  O  O  O  O  O
O  O  O  O  O  O  O  O  O  O  O  O  O  O  O  O
```

Input 1	1	1	O	O	O	1	O	1	1	1	1	1	1	O	1	O
Output 1	1	1	1	O	1	O	1	1	1	O	O	O	1	1	1	O
Input 2	O	1	O	O	1	1	O	1	1	O	O	O	O	1	1	O
Output 2	1	1	1	O	1	O	1	1	1	O	O	O	1	1	1	O
Input 3	1	O	O	O	1	O	1	1	O	O	O	O	1	1	O	O
Output 3	1	1	1	O	1	O	1	1	1	O	O	O	1	1	1	O
Input 4	O	O	O	1	1	1	1	1	O	1	1	1	O	O	1	O
Output 4	1	1	1	O	1	O	1	1	1	O	O	O	1	1	1	O
Input 5	O	O	O	1	O	1	1	O	O	O	O	O	1	1	O	1
Output 5	1	1	1	O	1	O	1	1	1	O	O	O	1	1	1	O
Input 6	O	O	1	O	O	1	O	O	1	O	1	O	1	1	O	1
Output 6	1	1	1	O	1	O	1	1	1	O	O	O	1	1	1	O
Input 7	O	1	1	O	O	O	1	1	1	1	1	O	O	1	1	O
Output 7	1	1	1	O	1	O	1	1	1	O	O	O	1	1	1	O
Input 8	O	O	1	1	O	1	O	1	O	1	O	1	1	1	1	1
Output 8	1	1	1	O	1	O	1	1	1	O	O	O	1	1	1	O
Input 9	O	1	O	1	O	O	O	1	O	1	O	O	O	1	1	1
Output 9	O	1	1	O	O	O	1	1	O	O	O	O	O	1	1	O

Fig 5-13. Example run of program HEBB3P, seed 72873.

MEMORY STATE 2

```
0  0  0  0  1  0  0  0  1  0  0  0  1  0  0  0
0  0  1  0  0  0  1  1  0  0  0  0  0  1  1  0
0  1  0  0  0  0  1  1  0  0  0  0  0  1  1  0
0  0  0  0  0  0  0  0  0  0  0  0  0  0  0  0
1  0  0  0  0  0  0  0  1  0  0  0  1  0  0  0
0  0  0  0  0  0  0  0  0  0  0  0  0  0  0  0
0  1  1  0  0  0  0  1  0  0  0  0  0  1  1  0
0  1  1  0  0  0  1  0  0  0  0  0  0  1  1  0
1  0  0  0  1  0  0  0  0  0  0  0  1  0  0  0
0  0  0  0  0  0  0  0  0  0  0  0  0  0  0  0
0  0  0  0  0  0  0  0  0  0  0  0  0  0  0  0
0  0  0  0  0  0  0  0  0  0  0  0  0  0  0  0
1  0  0  0  1  0  0  0  1  0  0  0  0  0  0  0
0  1  1  0  0  0  1  1  0  0  0  0  0  0  1  0
0  1  1  0  0  0  1  1  0  0  0  0  0  1  0  0
0  0  0  0  0  0  0  0  0  0  0  0  0  0  0  0
```

Input 10	1	0	1	1	0	1	1	0	0	1	1	1	1	1	1	0
Output 10	1	1	1	0	1	0	1	1	1	0	0	0	1	1	1	0
Input 11	1	1	0	0	0	0	0	1	1	0	1	0	0	0	0	0
Output 11	1	1	1	0	1	0	1	1	1	0	0	0	1	1	1	0
Input 12	1	1	0	1	0	0	1	1	1	0	1	0	0	0	1	1
Output 12	1	1	1	0	1	0	1	1	1	0	0	0	1	1	1	0
Input 13	1	1	1	1	1	1	1	0	0	1	0	0	0	0	0	1
Output 13	1	1	1	0	1	0	1	1	1	0	0	0	1	1	1	0
Input 14	1	0	1	0	1	0	0	0	0	0	1	1	0	0	1	1
Output 14	1	1	1	0	1	0	1	1	1	0	0	0	1	1	1	0
Input 15	1	1	0	1	0	0	1	1	1	0	1	0	0	1	0	1
Output 15	1	1	1	0	1	0	1	1	1	0	0	0	1	1	1	0
Input 16	1	1	1	1	0	1	0	0	1	1	1	0	0	0	1	0
Output 16	1	1	1	0	1	0	1	1	1	0	0	0	1	1	1	0
Input 17	1	1	1	0	1	1	0	1	0	1	0	0	1	0	1	0
Output 17	1	1	1	0	1	0	1	1	1	0	0	0	1	1	1	0
Input 18	1	1	1	0	1	0	1	0	0	0	0	0	0	1	1	0
Output 18	1	1	1	0	1	0	1	1	1	0	0	0	1	1	1	0

Fig. 5-13. (Continued)

Input 19	1	1	1	1	0	1	0	0	1	0	1	0	0	1	0	0
Output 19	1	1	1	0	1	0	1	1	1	0	0	0	1	1	1	0
Input 20	1	1	1	1	0	0	1	1	1	1	1	0	0	0	1	1
Output 20	1	1	1	0	1	0	1	1	1	0	0	0	1	1	1	0
Input 21	1	0	0	1	1	1	0	0	1	1	0	1	0	0	0	0
Output 21	1	0	0	0	1	0	0	0	1	0	0	0	1	0	0	0
Input 22	1	0	1	0	0	0	0	1	1	1	1	0	1	0	1	0
Output 22	1	1	1	0	1	0	1	1	1	0	0	0	1	1	1	0
Input 23	1	0	1	1	0	0	1	0	1	1	0	1	0	1	1	0
Output 23	1	1	1	0	1	0	1	1	1	0	0	0	1	1	1	0
Input 24	1	1	1	1	1	0	1	1	1	1	1	1	0	0	0	
Output 24	1	1	1	0	1	0	1	1	1	0	0	0	1	1	1	0
Input 25	1	0	0	0	1	1	1	1	1	1	1	0	1	1	1	0
Output 25	1	1	1	0	1	0	1	1	1	0	0	0	1	1	1	0
Input 26	1	0	1	0	1	1	1	1	1	0	0	1	1	1	1	0
Output 26	1	1	1	0	1	0	1	1	1	0	0	0	1	1	1	0
Input 27	1	0	0	0	0	1	0	1	0	0	0	0	0	1	0	1
Output 27	1	1	1	0	1	0	1	1	1	0	0	0	1	1	1	0
Input 28	0	1	1	0	0	0	1	1	0	0	0	0	1	0	1	0
Output 28	1	1	1	0	1	0	1	1	1	0	0	0	1	1	1	0
Input 29	1	1	1	0	0	0	0	1	0	1	0	1	1	0	0	1
Output 29	1	1	1	0	1	0	1	1	1	0	0	0	1	1	1	0
Input 30	0	1	0	0	0	0	0	1	1	1	0	0	0	0	1	0
Output 30	1	1	1	0	1	0	1	1	1	0	0	0	1	1	1	0
Input 31	1	1	0	1	0	0	0	1	0	0	0	0	0	0	1	1
Output 31	1	1	1	0	1	0	1	1	1	0	0	0	1	1	1	0
Input 32	1	1	1	0	1	1	1	0	1	0	1	0	1	0	1	1
Output 32	1	1	1	0	1	0	1	1	1	0	0	0	1	1	1	0

Fig. 5-13.   (Continued)

Input 33	0	1	1	1	0	1	0	0	0	1	0	0	0	1	0	0
Output 33	0	1	1	0	0	0	1	1	0	0	0	0	0	1	1	0
Input 34	1	0	1	0	0	1	0	0	0	0	1	0	0	1	1	0
Output 34	1	1	1	0	1	0	1	1	1	0	0	0	1	1	1	0
Input 35	0	1	0	1	1	0	0	0	0	0	0	1	0	1	0	0
Output 35	1	1	1	0	1	0	1	1	1	0	0	0	1	1	1	0
Input 36	1	0	0	0	0	1	1	0	0	0	0	1	1	1	0	0
Output 36	1	1	1	0	1	0	1	1	1	0	0	0	1	1	1	0
Input 37	1	0	0	1	0	0	0	1	0	1	0	0	0	0	1	1
Output 37	1	1	1	0	1	0	1	1	1	0	0	0	1	1	1	0
Input 38	1	1	1	1	1	0	0	1	0	0	0	0	0	0	0	1
Output 38	1	1	1	0	1	0	1	1	1	0	0	0	1	1	1	0
Input 39	1	1	1	1	1	0	1	0	1	1	0	1	0	1	0	1
Output 39	1	1	1	0	1	0	1	1	1	0	0	0	1	1	1	0
Input 40	0	1	1	0	0	0	0	1	1	1	1	1	0	0	0	1
Output 40	1	1	1	0	1	0	1	1	1	0	0	0	1	1	1	0
Input 41	0	1	1	0	0	1	0	1	1	0	0	0	1	1	0	1
Output 41	1	1	1	0	1	0	1	1	1	0	0	0	1	1	1	0
Input 42	0	0	1	0	0	0	0	0	1	0	1	0	1	0	0	0
Output 42	1	1	1	0	1	0	1	1	1	0	0	0	1	1	1	0
	0	0	0	1	1	1	1	1	0	1	1	1	1	0	0	0

Fig. 5-13. (Continued)

These spurious states can be caused by overlapping vectors in Hamming space. In other words, there are too many interconnections among the neurons. Some interesting programming experiments would be to include energy calculations between each iteration and see if there are, in fact, deeper stable states than the stored memory states. Another experiment would be a study of Hamming distance to see how far off one can be in Hamming space and still "pull in" to one of the stored states.

SUMMARY

In this chapter, I have discussed artificial neural networks. I have shown that stable attractor points arise in content-addressing and that simple limit cycles can also arise. The simulations included in this chapter are a type of cellular automata that consider *all* the cells, not just the nearest neighbor sites. The neurons were considered to be binary threshold logic devices.

Little (1974) has shown the existence of persistent states in the brain. Several other workers have discussed neural dynamics. Scott (1977) has written a landmark paper that is now somewhat dated. Harth (1983) and Babcock and Westervelt (1986) have written extensively about chaos and instabilities in neural networks. For readers interested in neural networks, I recommend the Snowbird conference proceedings edited by Denker (1986) and for a good introduction to the subject, see my earlier book (Rietman, 1988).

6

Julia Sets and Fractals

This chapter discusses iteration in the complex plain to study the Julia sets and the Mandelbrot set. Mandelbrot (1977) coined the term *fractal* for the geometric objects generated by iteration of the following equation:

$$Z_{n+1} = Z_n^2 + c$$

In the complex plain you have

$$Z = X + iY$$

where X and Y are the real and imaginary parts of the complex number Z.

JULIA SETS

The primary thrust of this chapter is devoted to a discussion of Julia Sets. The complex number

$$Z = X + iY$$

can be plotted on a complex plain where the X coordinate is the real axis and the Y coordinate is the imaginary axis. Using the iteration relation

$$Z_{n+1} = Z_n^2$$

there are only two attracting points. For an initial Z less than one, the attractor point is zero. For an initial Z greater than one, the attractor point is infinity.

If a complex constant is added at each iteration such that

$$Z_{n+1} = Z_n^2 + c$$

then the attracting points map out Julia sets and fractals. When the iterated points are plotted in Z-space and the parameter c is held fixed, the function maps out Julia sets. When the iterated points are started at $Z_o = 0$ and iterated for various values of the parameter c and plotted in c-space, the function maps out the Mandelbrot set. This can be made clearer with some algebra.

The basic equation is

$$Z_{n+1} = Z_n^2 + c$$

where Z and c are complex numbers given as follows:

$$Z = X + iY$$

$$c = P + iQ$$

The function is then the following:

$$X_{n+1} = X_n^2 + Y_n^2 + p$$

$$Y_{n+1} = 2X_nY_n + q$$

For Julia sets, you hold P and Q constant for the entire region of space, select an initial point (X_o, Y_o) and iterate this to an attractor point. The number of iterations required to reach the attractor point is recorded and assigned to a color. The point (X_o, Y_o) is then assigned this color and plotted. The Julia set is then a map of the number of iterations to reach an attractor point.

Many points do not reach an attractor point other than infinity. In order to prevent this, the modulus of the complex number must be calculated. Peitgen, et al. (1984) and Peitgen and Richter (1986) have shown that if the modulus is greater than two, the iterates will escape to infinity. The modulus is calculated by the following relation:

$$\text{mod}(Z) = \left| \text{sqr}(X^2 + Y^2) \right|$$

or

$$X^2 + Y^2 >= 4$$

The Julia sets are an infinite number of mappings for a whole range of constant values for P and Q.

The Mandelbrot set is only one set, but it is a very large set. Peitegen and Richter (1986) have shown many regions of the Mandelbrot set and Julia sets. The Mandelbrot set is found by selecting a value for the parameters P and Q, and holding these fixed while $Z_o = 0$ is iterated. The number of iterates needed to reach an attractor point or escape to infinity is assigned a color. The (P,Q) point is then plotted in c-space with the color attached to that (P,Q) point. Then another (P,Q) point is selected, $Z_o = 0$ is iterated again, and the process is repeated.

To reiterate: Julia sets are maps of Z-space and the Mandelbrot set is a map of c-space. The next section describes a computer program for Julia sets. The Mandelbrot set is not covered because it has been covered in the secondary literature quite well. I suggest Dewdney (1985), Dewdney (1987) and Fogg (1988) for more details on the Mandelbrot set.

ALGORITHM FOR JULIA SETS

The equation for Julia sets is given in the previous section. This section gives a line by line description of the program JULIA. The program, written in GWBASIC, should run on any computer after modification of the lines for file storage. The program begins in line 20 asking for a file name. Then lines 30 and 40 ask for the parameter P and Q. In line 50, a variable called KOUNT is initialized to zero. In line 60, the file is opened. Line 70 to 140 initialize variables.

JULIA

```
10        CLS
20        INPUT "INPUT FILE NAME ";FILE$
30        INPUT "INPUT P ";P
40        INPUT "INPUT Q ";Q
50         KOUNT=0
60         OPEN "O",#1,FILE$
70        XMAX=1.5
80        XMIN=-1.5
90        YMAX=1.5
100        YMIN=-1.5
110        DELTAX=.05
120        DELTAY=.05
130        MAXROW=(XMAX-XMIN)/DELTAX
140       MAXCOL=(YMAX-YMIN)/DELTAY
150       FOR COL = 0 TO MAXROW-1
160       FOR ROW = 0 TO MAXCOL-1
170       MODULUS=0!
```

```
180        CO=0
190        XLAST=XMIN+COL*DELTAX
200        YLAST=YMIN+ROW*DELTAY
210        WHILE (MODULUS < 4) AND (CO < 100)
220              XCUR=XLAST^2-YLAST^2+P
230              YCUR=2*XLAST*YLAST+Q
240              CO=CO+1
250              XLAST=XCUR
260              YLAST=YCUR
270              MODULUS=XCUR^2+YCUR^2
280        WEND
290        GOSUB 360
300        NEXT ROW
310        PRINT COL
320        NEXT COL
330        PRINT KOUNT
340        CLOSE #1
350        STOP
360        IF CO=100 THEN 390
370        PRINT #1,COL,ROW
380        KOUNT=KOUNT+1
390        RETURN
400        END
```

Loops begin in lines 150 and 160 to select row and column coordinates for what amounts to the Z initial value in 190 and 200. A WHILE loop begins in line 210 and ends in line 280. The WHILE loop is the iteration calculation to check the number of iterations before the attractor point or infinity is reached. In line 290, there is a GOSUB to print the data to the file. Line 360 selects only those points that reached an attractor point in less than 100 iterations to be printed to the data file. A RETURN in line 390 goes back to line 300 to select another coordinate point to iterate. After the program reaches the STOP in line 350, the data can be plotted using the program PLOT1.

Now let us look at some figures of Julia sets prepared by the program JULIA and plotted with the program PLOT1. The first thing to notice is that the Julia set is the region of the map not covered by data points. Because the data points printed to the file were those with less than 100 iterations before reaching infinity, these are the same data points plotted in the figures. If you chose to plot only those points that reached 100 iterations, the map would be a reversed image. In fact, it would be best to choose only those points that reach 100 iterations and to cover the space in finer increments to give better resolution to the picture. These figures took about thirty minutes to compute. Notice that Figs. 6-1 to 6-6 are connected

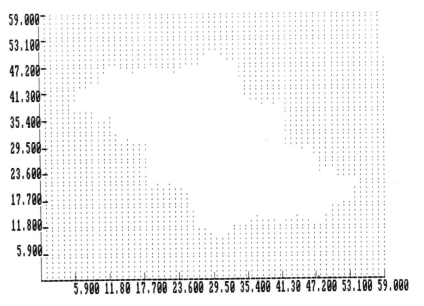

Fig. 6-1. Julia set P = −0.123, Q = 0.565.

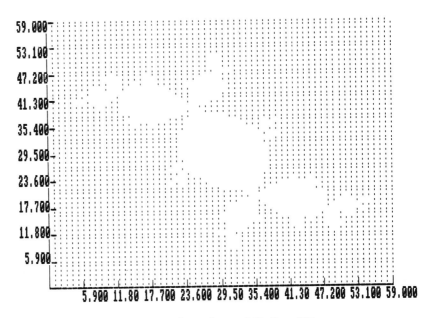

Fig. 6-2. Julia set P = −0.12, Q = 0.74.

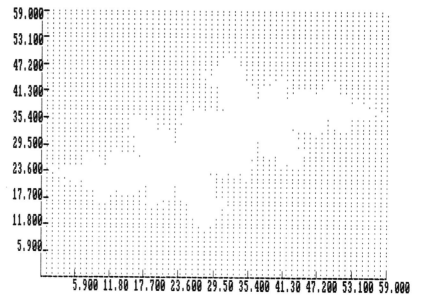

Fig. 6-3.   Julia set P = −0.4817, Q = −0.5316.

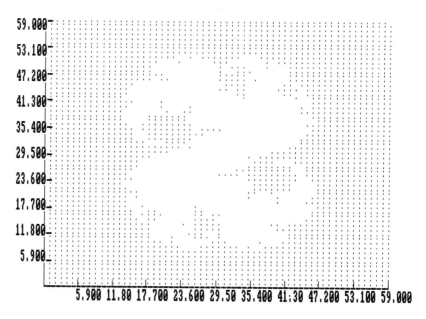

Fig. 6-4.   Julia set P = 0.2733, Q = 0.00742.

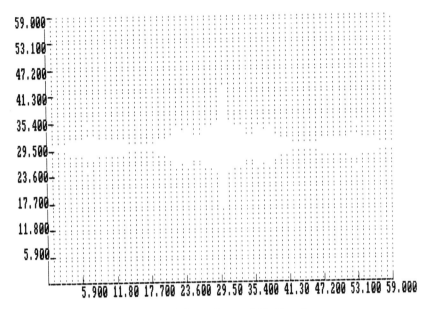

Fig. 6-5.   Julia set P = −1.25, Q = 0.

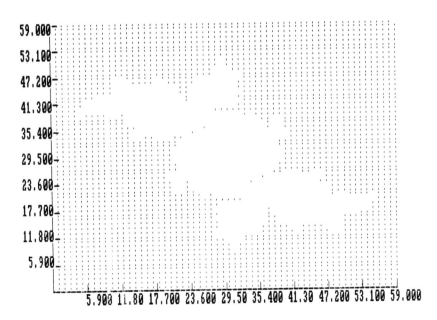

Fig. 6-6.   Julia set P = −0.11, Q = .6557.

Julia sets, while Figs. 6-7 to 6-9 are disconnected to various degrees. Some Julia sets are almost Cantor dusts. All Julia sets are fractals.

By using a color monitor and assigning colors to data points that take a certain number of iterations to reach a modulus of four, very beautiful figures can be created. For example, data points that take ten iterations to reach the modulus four could be assigned to a color 1. Points that take 11⁻20 iterations to reach modulus four could be assigned to a color 2, etc. The net result would be very beautiful figures of Julia sets. All Julia sets are fractals. Not all parts of the Mandelbrot set are fractals. Also note that the Mandelbrot set is always a connected set, but as I pointed out earlier, some Julia sets resemble Cantor dusts because they are so disconnected.

The Mandelbrot set, Julia sets, and fractals have been covered in the literature to such an extensive degree I see no point in belaboring the subject. For interested readers, I recommend the book by Mandelbrot (1977) and by Peitgen and Richter (1986). Applications of fractals to study dynamical systems have been covered in all the earlier chapters in this book. Applications of fractals to physics are covered by Pietronero and Tosatti (1985). Sander (1987) has discussed fractal growth models of crystals and air bubble movement in fluids. Orbach (1986) has discussed the dynamics of fractal networks in amorphous or glassy materials. Arcangelis

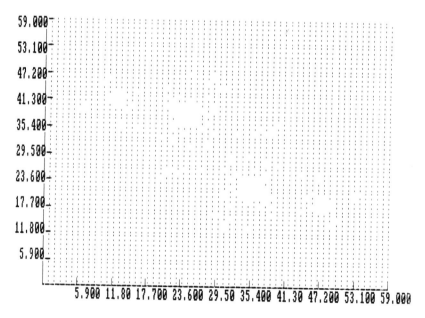

Fig. 6-7. Julia set P = −0.194, Q = 0.6557.

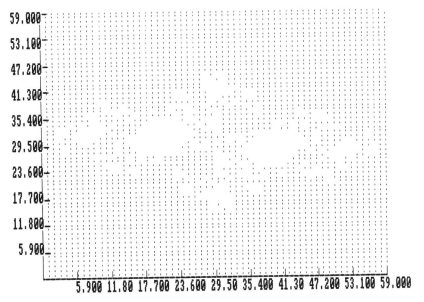

Fig. 6-8.   Julia set P = −0.74534, Q = −1.1301.

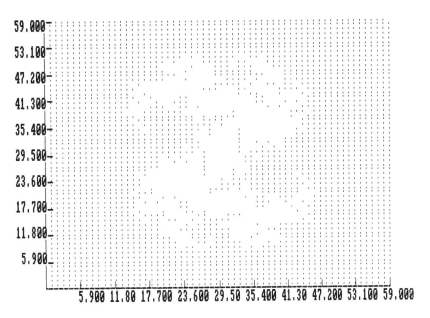

Fig. 6-9.   Julia set P = 0.32, Q = 0.043.

(1987) has discussed fractals in three-dimensional cellular automata. Mehaute (1985) has used fractals to discuss ionic transport in polymer-salt complexes. Sapoval, et al. (1986) have discussed the dynamics and creation of fractal objects by diffusion, and Bomes (1987) has shown that crumpled paper balls are fractal objects.

The list of fractal objects and applications for fractals is far too lengthy to review here. Again, I refer the interested reader to the literature.

7

Order Out of Chaos

In Chapter 3 you saw that strange attractors are information-producing systems Chapters 4 and 5 showed that cellular automata evolution can result in self-organized states. This chapter discusses self-organizing systems at greater length.

NONLINEAR NONEQUILIBRIUM THERMODYNAMICS

The main description of self-organization involves a mathematical description of nonlinear nonequilibrium thermodynamics. *Entropy*, the Law of disorder, is the starting point for the mathematics of self-organizing systems. Just as you saw in earlier chapters, such systems are energy-dissipating systems.

Entropy production is given by the following relation:

$$\sigma = \sum_i X_i J_i \geq 0$$

This states that the sum of the product of forces X_i and fluxes J_i must be positive. In a near-equilibrium state, the forces are weak and J_i may be expanded in a Taylor series:

$$J_i = J_i^{eq} + \sum_l \left(\frac{\partial J_i}{\partial X_l}\right) X_l + \ldots$$

At equilibrium, fluxes are zero so the following is true:

$$J_i = \sum_l \left(\frac{\partial J_i}{\partial X_l}\right)_{eq} X_l$$

The coefficients of proportionality

$$\left(\frac{\partial J_i}{\partial X_l}\right)_{eq} = L_{il}$$

are designated as phenomenological coefficients:

$$J_i = \sum_l L_{il} X_l$$

In 1931, Onsager showed that the following is true:

$$L_{il} = L_{li}$$

This relation reduced by one-half the number of coefficients to be determined experimentally. In the 1940s, Meixner and Prigogine devised a theory to explicitly calculate a number of these physical quantities. In the 1960s, DeGroot and Mazur (1962) applied the theory to many physical problems. Then Katchalsky (1965) applied the theory to biological systems in an attempt to discuss the nonequilibrium thermodynamics of living systems.

As an example of nonequilibrium nonlinear thermodynamics in bio-systems, consider the cell diagrammed in Fig. 7-1. The concentration of Na^+ ions outside the cell is far greater than inside the cell, and the concentration of K^+ ions is greater inside the cell than outside. By diffusion across the cell membrane, you would expect a flow of Na^+ ions into the cell and K^+ ions out of the cell. This, however, is not true. The membrane acts as an active pump to transport K^+ ions into the cell and Na^+ ions

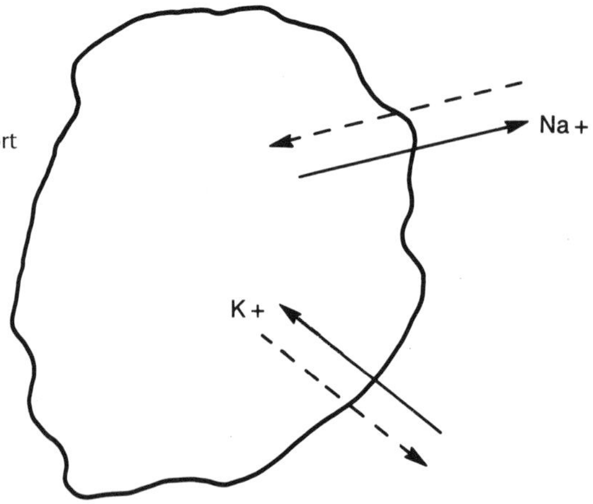

Fig. 7-1.   Diagram of ion transport in a cell.

out of the cell. The ions are in fact transported by the chemical reactions in the cell membrane.

There are many examples of nonlinear nonequilibrium thermodynamics in biosystems. Later I will discuss a few others, but now I would like to consider a physical system known as Benard convection.

BENARD CONVECTION

Benard convection is observed in almost all thermal and fluid interactions. For example, when milk is added to hot coffee, you can observe convection currents distributing the milk throughout the hot coffee. Benard convection is also the cause of some types of cloud formations in the atmosphere. This section gives a qualitative description of Benard convection.

Fig. 7-2. Schematic of Benard convection rolls viewed on edge.

Figure 7-2 represents Benard convection rolls. If a thin layer of viscous fluid at rest is heated slowly from the underside and the top surface is kept cool so the top surface is T_c and the lower surface is T_h, then the condition $T_h > T_c$ exists. This temperature difference over the layer thickness, h, is given by $(T_h - T_c)/h = \Delta T$. This gives rise to a constant heat flux, ΔT. As ΔT is slowly increased a critical value ΔT_c is reached in which the linear relation

$$J_i = \sum_l \left(\frac{\partial J_i}{\partial X_l} \right)_{eq} X_l$$

no longer holds. At ΔT_c, the fluid organizes to form convection rolls or convection cells of macroscopic size. The amazing aspect is that there is a correlation between molecules extending to distances on the order of a centimeter, and intermolecular forces extend only to distances on the order of 10^{-8}cm.

In Fig. 7-2, notice that adjacent rolls rotate in the opposite direction. This means that local symmetry is broken and the system is an energy-dissipative system. In order to have constant energy dissipation, there must be a constant input of energy to maintain stability of the convection rolls. Figure 7-3 is a photograph of Benard convection rolls in methyl alcohol. As ΔT is increased beyond ΔT_c, periodic behavior, quasiperiodic behavior,

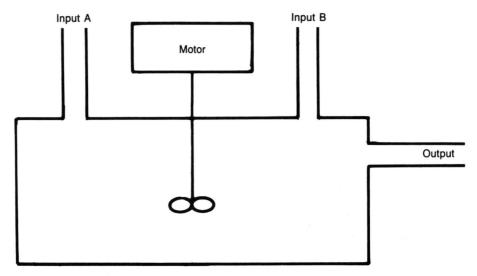

Fig. 7-3. Benard convection rolls in methol alcohol. Courtesy of Vincent Croquette E.N.S. Physique des Solids, Pairs.

and chaotic or turbulent motion results. The route to chaos appears to be a period-doubling route similar to that discussed for the logistic equation in Chapter 2.

BELOUSOV-ZHABOTINSKI REACTION

This section discusses chemical reactions that give rise to what are known as chemical clocks. These reactions are models for the clock cycles that occur in biological organisms.

As mentioned in Chapter 3, the Brusselator is a limit-cycle example. This example is a simplified model of a reaction known as the Belousov-Zhabotinski (B-Z) reaction. The relevant components of the reaction are as follows:

$$2BrO_3^- + 3CH_2(COOH)_2 + 2H^+ \longrightarrow$$
$$2BrCH(COOH)_2 + 3CO_2{\uparrow} + 4H_2O$$

There are over twenty intermediate steps to this reaction. This is a well-known chemical clock described by Babloyantz (1986), Field (1985), Epstein (1987), and Scott (1987). I mentioned the Belousov-Zhabotinski reaction to lead into the Brusselator, which is a two-variable model of the Belousov-Zhabotinski reaction.

There must be at least two variables to generate oscillations in a chemical system, $X = X(t)$, $Y = Y(t)$. Given a trimolecular step and auto-catalysis, there can then be a system that will exhibit chemical oscillations.

$$A \underset{k_{-1}}{\overset{k_1}{\rightleftharpoons}} X$$

$$B + X \underset{k_{-2}}{\overset{k_2}{\rightleftharpoons}} Y + D$$

$$2X + Y \underset{k_{-3}}{\overset{k_3}{\rightleftharpoons}} 3X$$

$$X \underset{k_{-4}}{\overset{k_4}{\rightleftharpoons}} E$$

The above scheme is the Brusselator. Assume that there is an open system similar to that sketched in Fig. 7-4. By adjustment of input fluxes J_a, J_b and output fluxes J_d, J_e the concentrations are held fixed for the components A, B, D, and E. These are the constraints of the system. The components X and Y are the response to the constraints. If we choose a unity for the rate constants, the Brusselator is given by the following system:

$$\frac{dx}{dt} = A - (B + 1)X + X^2Y$$

$$\frac{dy}{dt} = BX - X^2Y$$

If the system is held far from thermodynamic equilibrium, then time-dependent behavior becomes possible and chemical oscillations result.

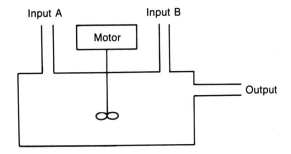

Fig. 7-4. Schematic of stirred reactor for B-Z reaction.

This system is modeled in Chapter 2 to describe limit cycles which are clear periodic systems. These chemical oscillators or chemical clocks are excellent examples of biochemical reactions in living systems. Rhythmic processes in living organisms, such as heartbeats and circadian rhythms, can be traced to biochemical clocks.

ORIGIN OF LIFE

There are many theories for the origin of life; all involve self-organization. Babloyantz (1986) discusses a very conventional theory. Kuppers (1985) discusses a molecular theory of evolution. Cairns-Smith (1985) discusses a very radical theory in which the first life forms had a crystal genetic code much like clay particles. The genetic code was overtaken by the polypeptides and polynucleotides of conventional life. Bienenstock, et al. (1986) edited a conference proceeding on disordered systems and biological systems. And Dawkins (1976, 1986) discusses biological organization and evolution in two very readable books.

THE GREAT RED SPOT OF JUPITER

Gleick (1987) discusses the work of Marcus, et al. (1985) on the formation of Jupiter's great red spot. Antipov, et al. (1986) show some results of an experiment to simulate the atmospheric conditions of Jupiter. There they show that the great red spot could be a *Rossby autosoliton*. This is basically a solitary standing vortex wave kept standing by counter-streaming zonal winds. At a high relative velocity, an instability is produced by the counter-streaming flows. This gives rise to a vortex or a self-organized system.

Self-organization from chaos is such an important topic that there is a new science called *synergetics* to cover the field. Haken (1977) wrote the first book on the subject and gave many examples of self-organization in physical, chemical, and biological systems.

Conclusion

Chaos is a hot new science that promises to explain some very complex phenomena. In this book I have tried to show a large number of examples where the new science of chaos is being applied. Similar mathematics describe subjects as diverse as developmental biology, galaxy formation, population dynamics, asteroid belts, atmospheric turbulance, microparticle formation, the origin of life, and neural dynamics, to name just a few. Dynamical equations for some of these phenomena (fluid turbulance, for example) have been known for centuries, but the equations could not be solved in a closed sense. With the advent of the computer, these systems can be approximated using discrete mathematics.

There is a great need for a description of these complex phenomena. The modeling and design of new aircraft could be improved with better understanding of fluid turbulance. The causes of rioting in the inner city could be predicted and perhaps dealt with before pressure built up to the point where a slight perturbation would be enough to trigger disaster. Phase-locking and period-doubling result in arrhythmia, or fibrillation of the heart—which is almost always fatal. A description of this could perhaps avert many untimely deaths.

The modeling of chaos can be very slow on conventional computers. For example, using the equations for turbulance, the weather can be predicted to a reasonable degree of accuracy for one day in advance, but the computation time is on the order of weeks.

New computer architectures are needed for faster computers. Many researchers are working on parallel processors, hypercube processors, mesh processors, pipeline processors, parallel analog computers, wafer scale integration, gallium arsenide, Josephen junctions, optical computers, and molecular scale processors. These new machines and ideas should speed up the development and understanding of the complex dynamics in chaos and self-organizing systems.

References

Antipow, S.V.; Nezlin, M.V.; Snezhkin E.N.; and Trobnikov A.S. *Nature* 323, 238, 1986.

Arcangelis, L. *Jour. Phys. A:Math* 20, L369-L373, 1987.

Arnold, V.I., *Russian Math.* Surveys 18, 9, 1963.

Babcock, K.L., and Westervelt, R.M. *Physica 23D*, 464, 1986.

————. *AIP Conference Proceedings #151*. Edited by J.S. Denker. New York: American Institute of Physics, 1986.

Babloyantz, A. *Molecules, Dynamics, and Life: An Introduction to Self-Organization of Matter.* New York: John Wiley Inc., 1986.

Bai-Lin, Hao. *Chaos.* Singapore: World Scientific, 1984.

Bak, P. *Physics Today,* 39, Dec. 1986.

Barenblatt, G.I.; Iooss, G.; Joseph, D.D. *Nonlinear Dynamics and Turbulence.* Pitman Publishing Company, 1983.

Batten, G.L. *Design and Application of Linear Computational Circuits.* Blue Ridge Summit: TAB BOOKS Inc., 1987.

Berge, P.; Pomeau, Y.; and Christian, V. *Order Within Chaos.* New York: John Wiley Inc., 1984.

Berlekamp, E.R.; Conway, J.H.; Guy, R.K., *Winning Ways for Your Mathematical Plays.* Vol.2: *Games in Particular.* Academic Press, 1982.

Bienenstock, E.; Soulie, F.F.; Weisknock, G. *Disordered Systems and Biological Organization.* New York: Springer-Verlag, 1986.

Boyce, W.E. and DiPrima, R.C. *Elementary Differential Equations*. New York: John Wiley Inc., 1977.

Cairns-Smith, A.G. *Seven Clues to the Origin of Life*. London: Cambridge University Press, 1985.

Carter, F. *Physica 10D*, 175, 1984.

Codd, E.F., *Cellular Automata*. New York: Academic Press, 1968.

Collet, P. and Eckmann, J. *Iterated Maps on the Interval as Dynamical Systems*. Boston: Birkhauser,1980.

Curry, J.H. *Comm. Math. Phys.* 68, 129, 1979.

Danby, J.M.A. *Computing Applications to Differential Equations*. Englewood Cliffs: Prentice-Hall, 1985.

Dawkins, R. *The Selfish Gene*. New York: Oxford University Press, 1976.

————. *The Blind Watchmaker*. New York: Norton & Company, 1987.

DeGroat, S., and Mazur, P. *Nonequilibrium Thermodynamics*. Amsterdam: North-Holland, 1962.

Denker, J.S. ed. *Neural Networks for Computing, AIP #151*. New York: American Institute of Physics, 1986.

Devany, R.L. *An Introduction to Chaotic Dynamical Systems*. Benjamin Commings Company, 1986.

Dewdney, A.K. *Scientific American*, May, 1985.

————. *Scientific American*, Aug. 1985

————. *Scientific American*, Nov. 1987.

Epstein, I.R. *Chemical & Engineering News*, March 30, 1987.

Farmer, J.D. *Physica 4D*, 366, 1982.

Feigenbaum, M.J. *Los Alamos Science*, Summer 1980.

Field, R.J. *American Scientist*, 73:142, March-April 1985.

Fogg, L. *Microcornucopia #39*, 6, Jan.-Feb. 1988.

Franceschini and Tebaldi. *Jour. of Stat. Physics* 21, 707, 1979.

Fredkin, E., and Toffoli, T. *Int. Jour. of Theor. Physics* 21, 219, 1982.

Freohling, H.; Crutchfield, J.P.; Farmer, D.; Packard, N.H.; Shaw, R. *Physica 3D*, 605, 1981.

Gleick, J. *Chaos: Making a New Science*. New York: Viking Press, 1987.

Goles, E., and Vichniac, G.Y. *Neural Networks for Computing, AIP #151*, Edited by J.S. Denker. New York: American Institute of Physics, 1986.

Gomes, M.A.F. *Jour. of Physics A:Math* 20, L283-L284, 1987.

Gould, H., and Tobachnik, J. *An Introduction to Computer Simulation Methods*. Vol. 1: *Application to Physical Systems*. New York: Addison-Wesley, 1988.

Gurel, D., and Gurel, O. *Oscillation in Chemical Reactions*. New York: Springer Verlag, 1983.

Haken, H. *Synergetics-An Introduction*. New York: Springer-Verlag, 1978.

Harth, E. *IEEE Trans. on Systems, Man and Cybernetics*, SMC-13 (5), 782-789, 1983.

Hayaski, Chihiro, *Nonlinear Oscillations in Physical Systems*. Princeton University Press, 1985.

Hebb, D.O. *The Organization of Behavior*. New York: John Wiley Inc., 1949.

Henon, M. *Commun. Math. Phys.* 50, 69, 1976.

————. *Quart. of Appl. Math.* 27, 291, 1969.

Hillis, W.D. *The Connection Machine*. MIT Press, 1985.

Holden, A.V. *Chaos*. Princeton University Press, 1986.

Hopfield, J.J. *Proc. Natl. Acad. Sci. USA* 79, 2554, 1982.

————. *Proc. Natl. Acad. Sci. USA* 81, 3088, 1984.

Hughes, G. *Byte,* 161, Dec. 1986.

Kalmogorov, A.N. *Dokl. Akad. Nauk. SSSR* 98, 527, 1954.

Kaneko, K. *Collapse of Tori and the Genesis of Chaos in Dissipative Systems*. Singapore: World Scientific, 1986.

Katchalsky, A., and Curran, P.F., *Noneqilibrium Thermodynamics in Biophysics*. Cambridge: Havard University Press, 1965.

Kuppers, B. *Molecular Theory of Evolution*. New York: Springer-Verlag, 1983.

Langton, C.G. *Physica 10D*, 135, 1984.

Lauwerier, H.A. "One-Dimensional Iterated Maps." *Chaos* edited by A.V. Holden. Princeton, 1986.

Lin, S-Y, and Lin, Y-F. *Set Theory: An Intuitive Approach*. Boston: Houghton Mifflin Company, 1974.

Little, W.A. *Math. Biosci.* 19, 101, 1974.

Lorenz, E. *Jour. of Atmospheric Sci.* 200, 130, 1963.

Mandelbrot, B.B. *The Fractal Geometry of Nature*. San Francisco: W.H. Freeman, 1977.

Marcus, Ingersoll, and Williams. *Proceeding of 110th Meeting of the Acoustical Soc. of America*, 1985.

Margolus, N. *Physica 10D*, 81, 1984.

Maxwell, L. and Reed, M. *The Theory of Graphs: A Basis for Network Theory*. New York: Pergamon Press, 1971.

May, R.M. *Nature* 261, 459, 1976.

McEliece, R.J.; Posner, E.C.; and Roderich, E.R., *Twenty-Third Annual Allerton Conference on Communication, Control, and Computing*, Oct. 1985.

Mehaute, A. *RISO International Symposium on Materials Science, Transport-Structure Relation of Fast Ion and Mixed Conductors*, 25-50, 1985.

Moser, J. *Nachr. Akad. Wiss. Gottingen, II Math. Physik K1*, 1, 1962.

Orbach, R. *Science* 231, 814, 1986.

Packard, N.H. "Complexity of Growing Patterns in Cellular Automata." *Dynamical Systems and Cellular Automata*, edited by Demongeot, Goles, and Tchoente. Academic Press, 1985.

Peitgen, H.O.; Saupe, D.; Haeseler, F.V. *Mathematical Intelligencer* 6(2), 11, 1984.

Peitgen, H.O., and Richter, P.H. *The Beauty of Fractals: Images of Complex Dynamical Systems*. New York: Springer-Verlag, 1986.

Pietronero, L., and Tosatti, E., eds. *Proceedings of the Sixth International Symposium on Fractals in Physics*. Amsterdam: North-Holland, 1985.

Potter, D. *Computational Physics*. New York: John Wiley, 1973.

Poundstone, W. *The Recursive Universe: Cosmic Complexity and the Limits of Scientific Knowledge*. Chicago: Contempory Books, 1985.

Rietman, E.A. *Experiments in Artifical Neural Networks*. Blue Ridge Summit: TAB BOOKS, 1988.

Robbins, K.A. *SIAM Jour. of Appl. Math* 36, 457, 1979.

Robert, F. *Discrete Iterations: A Metric Study*. New York: Springer-Verlag, 1986.

Rossler, O.E. *Z. Naturforsch* 31A, 1664, 1976.

Ruelle, D. *Mathematical Intellegencer* 2(3), 126, 1980.

Rumelhart, D.E., McClelland, J.L., and the PDP Research Group. *Parallel Distributed Processing: Explorations in the Microstructure of Cognation*. Vol. 1: *Foundations*. MIT Press, 1986.

Sander, L.M. *Scientific American*, 94, Jan., 1987.

Sapoval, B.; Rosso, M.; Gouyet, J.; Colonna, J. *Solid State Ionics*, 18 & 19, 21-30, 1986.

Scott, A.C. *Jour. of Math. Psychology* 15, 1-45, 1977.

Shoup, T.E. *Numerical Methods of the Personal Computer*. Englewood Cliffs: Prentice-Hall, 1983.

Sparrow, C. *The Lorenz Equation: Bifurcations, Chaos, and Strange Attractors*. New York: Springer-Verlag, 1982.

Tedeschini-Lallin, L. *Jour. of Stat. Physics* 27, 365, 1982.

Thompson, J.M.T., and Thompson, R.J. *The Inst. of Math. and Its Appl.* 16, 150, April 1980.

Toffoli, T. *Physica 10D*, 117, 1984.

————. *Physica 10D*, 195, 1984.

Van der Pol, B. *Phil. Mag.* 7-2, 978, 1926.

————. *Phil. Mag.* 7-3, 65, 1927.

Vannimenus, J.; Nadal, J.P.; Derrida, B. "Stochastic Models of Cluster Growth." *Dynamical Systems and Cellular Automata*, edited by Demongeot, Goles, and Tchneute. Academic Press, 1985.

Vichniac, G.Y. *Physica 10D*, 96, 1984.

von Neuman, J. *Theory of Self-Reproducing Automata*, edited by W.A. Burke. University of Illinois Press, 1966.

Wolfram, S. *Rev. Mod. Phys.* 55 (3), 601, 1983.

————. "Some Recent Results and Questions about Cellular Automata." *Dynamical Systems and Cellular Automata*, edited by Demongeot, Goles, and Tchneute. Academic Press, 1985.

Index

The Advanced Programming Technology Series

Exploring Natural Language Processing:
Writing BASIC Programs that Understand English

by David Leithauser

Exploring the Geometry of Nature:
Computer Modeling of Chaos, Fractals, Cellular Automata and Neural Networks

by Edward Reitman